Fables and Futures

Biotechnology, Disability, and the Stories We Tell Ourselves

George Estreich

The MIT Press
Cambridge, Massachusetts
London, England

This book was set in Stone Serif by Westchester Publishing Services. Printed and bound in the United States of America.

Library of Congress Cataloging-in-Publication Data

Names: Estreich, George, author.
Title: Fables and futures : biotechnology, disability, and the stories we tell ourselves / George Estreich.
Description: Cambridge, MA : The MIT Press, [2019] | Includes bibliographical references and index.
Identifiers: LCCN 2018027592 | ISBN 9780262039567 (hardcover : alk. paper)
Subjects: LCSH: Biotechnology. | Human genetics. | Parents of children with disabilities.
Classification: LCC TP248.2 .E83 2019 | DDC 660.6--dc23 LC record available at https://lccn.loc.gov/2018027592

10 9 8 7 6 5 4 3 2 1

in memory of my father
Paul J. Estreich
1930–1996

Technology is neither good nor bad; nor is it neutral.
—Melvin Kranzberg

A world where everyone is welcome is the first priority.
—Marsha Saxton

Contents

Introduction

In late March of 2001, a month after my younger daughter Laura was born, then diagnosed with Down syndrome, I drove out to the Oregon desert. Theresa, my wife, had taken Laura and her older sister Ellie back East, to visit family; I wanted some distance, a change of landscape. A broader perspective, or at least a new one. I was with a friend I'll call Robert, the first person I told about Laura. In the weeks since Laura's diagnosis, he'd been listening more than talking; when he spoke, he was measured, considerate. He had distance without being distant. For me and Theresa, mired in simple yet unassimilable facts—Laura's extra chromosome, her condition, her heart defect, as yet unrepaired—Robert's presence was as consoling as any empathetic shock.

We loaded up tents, packs, ramen noodles, and candy bars. Then we shooed the two dogs into the back of the Subaru: his dog, an expressive, jowly English pointer, and Penny, our anxious, submissive, cinnamon-colored mutt, fattened by years of food falling from high chairs. They scrabbled and slobbered and whined. We set out, calling back to them to calm down.

It felt good to be driving east, toward the usual limits of vision. We live in the Willamette Valley, a wide north–south swath of agricultural grass fields, bounded to the west by the low, softened mountains of the Coast Range and to the east by the intermittent snowy peaks of the Cascades: Mount Jefferson, Three Fingered Jack, the Three Sisters, widely spaced as punctuation for an unseen sentence. As you descend from the pass, the Douglas firs recede, replaced by ponderosa pines, then juniper, then sage; the ground, once loamy, is cindery and dry; the air smells different, rarefied, sharper. Soon it's all fence wire, sky, and the scoured forms of hills, their cinnamons and russets knotted by sagebrush, like the back of a rug.

As the landscape emptied out, I began talking about Laura, venting, the words spilling over. I didn't know how to think about her. I remember saying I wanted her to be *Laura*, my daughter, not a medical case, not an extra chromosome. I could already hear the iron doors of categories clanging shut, *special needs, disabled, developmentally delayed.* I didn't want her to be the sum of her medical problems. I wanted some sliver of her life left pure. Robert mostly listened and kept his counsel. I do remember his saying that Down syndrome seemed to him part of the normal range of human variation, and that, as such, Laura might be unique and valuable not in spite of her Down syndrome, but because of it.

We arrived late, switchbacking up to the plateau where the campground was, and set up tents in the shadow of rounded hills, hurrying to feed the dogs and ourselves before dark. In the evening, we walked up the dirt road from the campsite. The dogs bounded ahead, researching invisible trails.

You could still see some orange in the west. A few bright clouds were ashing over. Then the first star was out, and the sky was all gradations of blue. But one cloud stayed, pink as the inside of a grapefruit. It was different. The other clouds were, definably, *volumes*, objects in the sky. A child could look up and say Car, or Horse. But this one had no shape you could name by likeness. It was a blur, a region of unfocus, a mist. It was, we thought, simply higher than the other clouds, thinnest cirrus gathering the last of a sun we had already turned away from. But as the sky darkened, the cloud's colors grew deeper, more saturated. It seemed lit from within. We could see the light moving, rippling with soft verticals, like folds in a heavy curtain. A dark coruscation, a pulse.

It was clear that we were looking, somehow, at the aurora borealis: a rare event at this latitude, but evidently there. The cloud was not a cloud. It had been right in front of us all along, but expectation had kept us from seeing it for what it was.

* * *

I was trying to answer the question that disability had posed. I was doing so in a world that abounds with wrong answers, that did not even have satisfactory language with which to begin to speak.

This book explores our ongoing American conversation about human-focused biotechnology: applications able to "read" and "write" DNA, the molecule of which genes are composed. If I begin with a story about Laura, it's not only because her condition is genetic; it's because stories about disability are central to that conversation. Even as I drove out to the desert, I was beginning to understand the shape of the standard story, to understand how much of it I had already inhaled and made my own. I was discovering the fables I already believed. In the years afterward, as I worked on a book about Laura, I learned and relearned the obvious but vital truth that narrative questions are also political ones: who gets a story, who gets to tell it, whose stories are credited, and the limits of stories themselves.

In March of 2001, I was submerged in shock, but my approximate understanding of DNA and chromosomes had improved. DNA is the long, double-stranded molecule necessary for life. It contains the instructions to build, maintain, and replicate an organism. DNA's structure, the familiar double helix, is unchanging, but the sequence of its paired component bases—adenine, thymine, guanine, and cytosine, abbreviated A, T, G, and C—is variable: differences in life-forms (plane trees, rhinoceroses, E. coli, semi-employed writers) begin as differences in sequence. Significant stretches of sequence are called *genes,* which mainly encode the complex structures called proteins, on which life depends. Humans have between 20,000 and 25,000 genes, though these constitute only a fraction of the total DNA we inherit, which is called the *genome*. In humans, the long thread of the genome is bundled into *chromosomes*. Most of us have forty-six: twenty-two pairs of autosomes, plus the two sex chromosomes, XX for women, XY for men. People with Down syndrome have forty-seven, with an extra copy of the twenty-first chromosome, the smallest, containing around two hundred genes (*trisomy 21*: three copies of chromosome 21). Laura's apparent differences from us, it turned out, sprang from an abundance of similarity. These microscopic facts were joined to macroscopic mysteries, including Laura's appearance, her heart defect, and the loss of what I understood to be an ordinary life.

I did not even know how to think about Laura, she was both my actual daughter and a vague radiance in the Western sky—a phenomenon I could see but not name—and so I could barely begin to think about intellectual disability and society, let alone the complications of prenatal testing, human gene editing, virtual children, de-extincted woolly mammoths, or semi-synthetic cells. But as the strangeness faded, as Laura survived her heart surgery and a feeding disorder and began to thrive and emerge from the facts and fears in which I had cocooned her, I came to understand that the chromosome mattered less than its context. Our lives were entangled in something larger, in a long American conversation about genes and disability and difference and family and normality and belonging. That conversation is shaped, in part, by an awareness of biotechnology's capabilities. This was evident early in Laura's life, when every discussion of her arrival was marked by a question, spoken or unspoken, about whether we had tested. But questions about biotechnology are rapidly becoming questions for everyone.

We live at a unique time in history: we are able, as never before, to translate genetic information into digital information and back again. This means that we are increasingly able to predict the qualities of future people—and to alter them, should we choose to, in inheritable ways. Therefore, the technology implies necessary questions about what kinds of people we value, and about how the most vulnerable will be affected. Though many are thinking and talking about these subjects, the *popular* conversation is still tiny, considering the species-altering changes on offer. That's why the conversation matters so much, and because that conversation is pervaded by narrative, the stories matter too.

*   *   *

Every new technology is accompanied by a persuasive story, one that minimizes downsides and promises enormous benefits. I use the term "story" loosely: in our fragmented, digital time, the narrative is more often implied, composed of images as well as words. However, beneath the fables attached to any one application—the perfect and perfectly content families in ads for next-generation prenatal tests, the pristine landscapes with mammoths awakened from extinction's slumber—is a progress narrative, in which science and technology bring us a better world. Too often that narrative frames disability as a cost. I'm deeply pro-science, and I'm fond of advanced technology—from the computer on which I type these words to the ECMO machine that kept Laura alive while her heart was repaired. But if our new biotechnologies are to help us, we must see them clearly—and we must also recognize that, as currently constituted, they reflect and amplify our misconceptions about disability and our devotion to the often-destructive idea of "normal."

In the conversation about human-focused biotechnology, people with disabilities are mostly invisible; when present, they exist as emblems more than individuals, as occasions for ethical debate, or as examples of outcomes to avoid. Their identities are occluded by diagnosis or stereotype, their interiority goes unacknowledged and unremarked, their emotions are simplified by design, and they are rarely consulted. The paradox is that whether as absence or distortion, they are essential. As a rhetorical device, disability offers the rationale for the technology's development and use: that is, a central promise of the technology is that disability will be repaired or prevented.

This is not to say that disability is always a "good" thing being portrayed as "bad," but rather that disability is complex and flattened into simplicity. Key to that oversimplification is the view that "disability" is something purely physical, something comprehensible through science and medicine and ameliorable, or preventable, through technology. This approach—in the shorthand of scholars in disability studies, the "medical model of disability"—downplays the role of social context in creating disability's meaning, blurs key distinctions between disability and disease, and omits the perspectives of many with disabilities. For that reason, I have tried to include some of those voices here.

*   *   *

Before Laura came along, people with intellectual disabilities lived in the periphery of my vision. Laura's arrival thrust disability from the periphery to the center, and in response to that dislocation, I wrote a memoir. I had many reasons for doing so, but one was that since the attention Laura drew was inevitable, I might as well work with it. If people were going to stare, I might as well lend some depth to the picture. They were, as I found, often staring at a projection. Call it a huggable ghost: a vague shape, a diagnosis with a personality, a mix of sweetness and tragedy, of angels and heart defects and maternal age. That is a way of imagining Down syndrome, and not the worst way, but it hides the individual. The projection, the ghost, obscures the child.

This book had its genesis in that one. In writing the memoir, I came to understand that questions about Down syndrome are really questions about disability; that questions about disability are questions about who counts as human; that these questions are inextricable from our understanding of cutting-edge biotechnologies; and that, because few of us are experts, our understanding of new biotechnologies is shaped not only by the cultural assumptions we already hold, but also by instances of authoritative persuasion.

The strangeness of having a daughter with Down syndrome faded long ago. In its place is the strangeness of the world—or, more precisely, the strangeness of living in a world that cannot make up its mind about Down syndrome in particular or disability in general. I could resolve, or manage, my own conflicts. As for the world's, I can only describe them. It is as if my house were bisected by a border: The front door opens on one country, the back door on another. In one country Laura is valued and seen, and her difficulties are neither stigmatized, nor held against her, nor used to argue that others like her should not exist; they are accommodated, and she receives what she needs without grudge or pity. In the other country, she is an emblem of the failed pregnancy, a synonym for tragedy, a target of ridicule, an expense to the nation, and the opposite of progress. Each vision has its accompanying future. One is a vision of belonging, for people of all brains and bodies. The other is a vision of progress, in which high-tech directed evolution eliminates genetic disease, leading to health, enhanced abilities, even a new species. Though these visions are in tension, they are neither exclusive nor symmetrically opposed, and what will actually happen is impossible to know. Like most parents I am haunted by the future, but I know better than to try and predict it.

*   *   *

This book is the record of an exploration. I am a writer, a nonspecialist in a specialist's wilderness, and I find my way with a writer's tools: reading closely, thinking about what I read, and answering the stories I encounter with stories of my own. Even if I could claim a specialist's expertise—and I know enough actual experts, from scientists to historians to literary critics, not to try—a truly comprehensive, in-depth understanding would have to include a dozen-odd fields, including (for starters) the history of eugenics, disability theory, bioethics, epigenetics, rhetoric, and international law. No unenhanced human being can do that. Human-focused biotechnologies are information technologies; the humans are overwhelmed with information.

So we need to do the best we can. We should read up and come to our own conclusions, neither disregarding experts nor being cowed into silence. If we wait for total expertise, the technology will have moved on; and in any event, we are each expert in what it means to be human, to live in a body, to be one of a species whose germline has not yet, at this writing, been engineered. It is my hope, then, that one nonspecialist's exploration may be useful: that others will be encouraged to find their own ways through the wilderness, to come to their own conclusions, and to join the conversation this book examines. I have few certainties, but one is that people-changing technologies are more likely to be used wisely if more people talk about them.

A brief word, then, about the book's approach. Each chapter focuses on a single biotech application, using it as the occasion for meditating on some intersection of biotechnology, disability, and the way we talk about both. Even—or especially—when an explanation of technology is presented as information, it's key to remember that a description is not only accurate or inaccurate. It is a force in the world, a wave in the ocean of messages we live in. I'm interested in descriptions with force, the widespread accounts aimed at lay people: the accounts of experts, or those that have the sheen of expertise.

Many of these take the form of prediction—especially since the development of the CRISPR-Cas9 system, a fast, relatively precise way of "editing" the genes of living cells. CRISPR has already been used to alter embryos in vitro, and it may one day be used for reproduction, to avoid disease or enhance human traits. So predictions are understandable. But to quote the authors of "CRISPR Democracy: Gene Editing and the Need for Inclusive Deliberation," "[r]evolutionary moments do not reveal the future with map-like clarity":

> Far more, they are moments of provisionality, in which new horizons and previously foreclosed pathways become visible. The challenge for democracy and governance is to confront the unscripted future presented by technological advances and to guide it in ways that synchronize with democratically articulated visions of the good.

This book is written in a moment of provisionality, and it examines the stories we tell ourselves, the scripts we bring to the unscripted, the fables that help to create our futures. Given biotech that can select and shape who we are, we need to imagine, as broadly as possible, what it means to belong.

* * *

I haven't been to the Eastern Oregon desert in years. The dogs died long ago. The Subaru gave out and was hauled away to auction. Laura's in high school, and Ellie is a college graduate. I wrote my way from then to now, drafting and revising the story as it unfolded, which is what I was beginning to do as Robert and I drove across the state and I wondered what it meant to have a daughter with a disability. I know now that there is no final answer, that it is up to me and Theresa, and not only to us: to Ellie, to the world around us, to Laura herself. This is the long view Robert took, or what I took from what he said: that the extra chromosome did not have a definite meaning—tragic or otherwise—to grasp, but that it would take its meaning from what others thought and did and said; that the chromosome mattered less than the child herself; and that questions of meaning, particularly for a parent, were less important than the quality of a daughter's life. I think about Laura a lot. In the end, what matters is what she thinks about herself.

Looking back, I wish, a little bit, that I could have had more of this understanding then. That I could donate what wisdom I have to my younger self, like a kidney. But it is more like a face, something indistinguishable from identity and time, an understanding not only grasped, but earned. In a time of transformations—genes to data, data to words—that is the transformation I write from.

Everything I said to Robert, as we drove across the desert, was fueled by an inner argument: I believed that an ordinary happiness was possible; I believed it had been denied. But what counts as ordinary? What seemed a cloud became the Northern lights—extraordinary for me, ordinary somewhere else. It was a glimpse of life at another latitude, of an unfamiliar sky. I know now that the terms of my wishing were mistaken. Ordinary life had not been lost; it was always a mirage, like the idea of a normal child.

1 Virtual Children

Hi Dad good morning is 9:52

Hi Laura are you dressed yet?

Oh need to have one second

It's March. A sunny Saturday morning. I'm in the garden, clearing out last year's vines, planting onion sets, texting with Laura. She'll be out in a few minutes, asking me to fix her eggs and toast. She's perfectly capable and cooks herself breakfast all the time but generally would prefer that someone else do it. Another ordinary moment I could not have imagined, seventeen years ago, when I was driving through the Oregon desert and trying to imagine the future.

If I had known what was coming, I would have worried less. Laura's high school experience is atypically good. She's taking Spanish: the words that stick (*perro*, dog; *bailar*, to dance; *nadar*, to swim) are the ones that register with her life. She's on the varsity cheerleading team and plays basketball on a Special Olympics Unified Team that includes both disabled and non-disabled athletes. Nothing is perfect—when is it ever?—but Laura is happy, and her public school experience is so far free of the horror stories I hear from other parents near and far. This fragile bubble of more-than-okayness lasts as long as I don't sign onto the Internet, where *retard* is the comment-thread term of art across the political spectrum; where politer voices frame children with Down syndrome as cute but regrettable; and where disability, even for many of those with whom I vote, is a forgotten category and a synonym for contempt. I don't know what future will unfold from this tension or what place Laura will have in it. But I do think that future depends, in part, on what we think and say.

Raising a child with Down syndrome resembles writing science fiction: you are always trying to imagine, to write into existence, a world where your child truly belongs. The act is utopian, yet constrained by the possible. In my case, the story features a world where educational inclusion for children with disabilities is easy and assumed, where adults are able to work for a living wage, and where the sexual violence against people with intellectual disabilities is statistically insignificant. Laura is the reason I think about the things I do. If I care about the future, it is because I want her to have a place there; if I think about the stories we tell ourselves, it's because the fictions to which others subscribe are the force fields around her life; and if I write about family, it's because *family* is the arena where old fictions and new technologies collide.

* * *

In 1997, Princeton biologist Lee Silver published *Remaking Eden: How Genetic Engineering and Cloning Will Transform the American Family*. In it, Silver predicts a steady expansion of biotech intervention. Testing expands in scope, from diseases to desired traits. We move from selection to alteration, from reading genes to writing them, until a distant day when the human species is upgraded beyond recognition.

Silver advances his book's argument with brief speculative fictions. In one near-future scenario, "Melissa" and "Curtis" are trying to choose which of their eighty-four frozen embryos will be implanted. They already know they want a girl, and that the girl will be named Alice. Melissa is at a computer screen, sorting through embryos, each of which has a computer-generated future portrait. She examines risk scores—first for severe diseases, then for common diseases from heart disease to cancer, then for "Physiological and Physical Characteristics," then for individual abilities (like musical talent), then for "small aspects of physical appearance."

> Number forty-three had a chin, nose, and eyes that made her resemble her father, and this appealed to Curtis. But Melissa thought that number seventeen would turn out to be more pretty, although this highly subjective aspect of physical appearance would always be difficult to gauge in a computer-generated facial image.

Biotechnology is often advanced with idealized narratives, in which appealing families make use of reliable technologies. This is one such narrative, a fable of consumption. Both characters are nice. They only want the best. They get along. Melissa isn't a stage mom, and Curtis isn't an aging, failed ballplayer living through his child. Though Melissa wants "a daughter with an innate score for musical talent," and Curtis wants Alice "to be born with temperament and cognitive attributes that would serve her well in the business world," they agree about having a girl, and their final choice does not expose an unbridgeable rift between their values re: parenting or religion. They also understand "that there is no such thing as a perfect child," and they are able to compromise without conflict:

> The one with the strongest musical talent was on the shy side and lacking in talent for abstract analysis, which did not appeal to Curtis. Likewise, the one with the strongest talent in abstract analysis would have a tendency to be twenty pounds overweight and not very musical, which Melissa was unable to accept. So compromise was the word of the day.

Like the characters, the technology is presented in a positive light. In the scenario, knowing an embryo's genome means that complex human attributes can be reliably predicted—and simplified to a single number for purposes of comparison. Not only that, but these attributes can be finely distinguished. "Musical talent" is apparently separable from "abstract analysis," though what's more abstract than music? The exacting math of a Bach invention, or Thelonious Monk, calculating a run of chord substitutions in real time—are these not instances of "abstract analysis"? Can the musical kind really be distinguished from the business-friendly kind, let alone be predicted from gene variants? Can *any* complex, vaguely defined category be traced to something as specific as a sequence of nucleotides? In Silver's narrative, they can—and so can elements of personality. The last two choices "scored well on all important temperament scales with a tendency toward long-term happiness, emotional stability, conscientiousness, mild risk-taking behavior, and an assertive but not overaggressive nature together with an outgoing personality."

Melissa and Curtis make their choice, and "nine months later, they shared in the thrill of the birth of their baby girl—a real-life Alice." The story I want to read begins there, with Alice learning the specific wishes engraved in her genome, with adolescent Alice writing poetry alone in her room about being Alice Number Forty-Three, with Alice's arguments with her mother over body image and second helpings and how long to practice the piano, or with Alice's decision to quit her prebusiness major and apply her talents in abstract analysis for an indefinite period on an organic farm, or with Alice's struggle to define her own identity, her own story, against the intentions encoded in her before birth. Alice trying to figure it out, the extraordinary, fine-grained act of control written into her cells. But Silver's narrative steers clear of the complexities that spring from individual stories. It is basically a commercial for the future, in which two middle-class parents shop carefully for a child genetically adapted for future American success: outgoing, attractive, talented, business savvy. The story has a dystopian frisson, but like the rest of Silver's book, it is less cautionary than enthusiastic.

*   *   *

In *Remaking Eden*, Alice's story opens a chapter entitled "The Virtual Child." In 2009, Lee Silver cofounded a company named GenePeeks, whose business model depends on generating virtual children.

Using a proprietary algorithm, GenePeeks analyzes DNA samples from two people—a single woman, for example, and a candidate sperm donor—and then generates hypothetical children in silico to estimate the disease risk of a child produced by the match. Unlike the eighty-four embryos in the Alice scenario, the children have no biological reality. The spit or sperm samples on which they are based are abstracted into information: reduced to DNA sequence, converted to digital information, processed by an algorithm, retranslated into a risk score, a recommendation for action. The virtual children are passing artifacts of big data, immaterial beyond the company's servers, fictions in the machine. And yet they have the power to shape a specific future, to determine whether one Alice or another will be born. They are probabilistic ghosts, though like all ghosts, they are all the more powerful for being unreal.

In the company's early days, GenePeeks's home page had only two images: a beautiful, healthy baby and a laptop. Two double helixes, one red and one blue, streamed from above onto the computer. A genomic family portrait: the child alone, the parents represented as DNA, as genes becoming bytes, as one kind of information becoming another. An uncluttered white background suggested labs, medical spaces, and purity. Context was absent, leaving only the promise of biotechnology and an image of a perfect child. Around that time, *Science* profiled the company; the reporter, Jennifer Couzin-Frankel, wrote that Silver and cofounder Anne Morriss were "still sorting out which diseases to include in testing":

> "The initial approach is to focus on serious, life-changing, life-threatening childhood diseases," Morriss says. This might include hundreds of rare recessive diseases, "because there's no controversy" there, Silver says.

If we take the founders at their word, the list of candidate conditions was, in part, a business decision. In addition to the (broad) criteria of "serious, life-changing, life-threatening," the key factors were technology (conditions potentially calculable by algorithm) and public relations (the avoidance of controversy). In other words, the working definition of health, the idea translated into action by the algorithm, was a function of persuasion.

The discussion of biotechnology often centers on ethical questions, like when, if ever, inheritable genetic modification might be permissible or even desirable. These questions matter, but the actual choices will not occur in the rarefied atmosphere of bioethical discussion. They will come to us as products, which means that market forces will drive both which products are offered and the persuasion that accompanies them. Because the products need to answer to existing demand, they will reproduce the values of the society in which they are sold. To the extent that they are adopted, they will translate those values into human populations.

If what counted as "health" were obvious and unambiguous, then these questions would be footnotes to progress. But the long history of stigma attached to disability, and the rapid changes in our understanding of health and disease, suggest that the questions are central. Andrew Solomon, writing in *The Guardian*, notes that "scientific definitions of illness often run up against humanist definitions of identity" and continues:

> I know whereof I speak: I have an identity that was long deemed an illness. The literature on homosexuality as a disease is amply represented in medical libraries, and it is my own history, even though I disbelieve it. The process through which gay identity was rescued from medical textbooks launched humanism's upstaging of medicine. It reflects the common clinical presumption that variation from the norm constitutes pathology.

Two other points bear mentioning. The first is that, for a company like GenePeeks, market forces will tend to drive an expansion of the boundaries of health. Having a long list of conditions both makes the application seem more powerful and gives the consumer more to worry about. The second is that the idea of health is often used to leverage that very expansion. Disease offers the "initial approach," but disease may only be the beginning.

* * *

The feature article in *Science*, where the founders discussed which diseases they might test for, was published in October of 2012. But in August of that year, Silver had already filed for a patent on his algorithm. The patent—granted in late 2013—makes clear that from the beginning, Gene-Peeks's vision ranged well beyond disease.

Speaking to *New Scientist* in 2014, Silver explained that the patent "covers any disease or any trait that has a genetic influence," which is one way of saying "nearly any human feature." *Method and system for generating a virtual progeny genome, US 8620594 B2*, includes a very long list of diseases, but it also names the following traits: *ability to roll the tongue, addiction(s), aggression, blinking, breast size/shape, cleft chin, conditioned emotional response, dimples, discrimination learning, drinking behavior, drug abuse, eye color/shape, eyelash length, ejaculation function, face shape, grip strength, grooming behavior, hair color, hair texture, handedness, hearing function, height, learning/memory, lips size/shape, maternal/paternal behavior, mating patterns, mental acuity, mental stability, mental state, nose, odor preference, penile erection function, posture, propensity to cross the right thumb over the left thumb when clasping hands or vise* [sic] *versa, response to new environment (transfer arousal), self tolerance, senility, skin color, social intelligence, temporal memory, urination pattern, weakness, widows* [sic] *peak or lack thereof, etc.* The *etc.* is in the original list, which is much longer. According to the patent, the test can be "authorized by, referred by, and/or performed by" a host of professionals or professional organizations, including labs, genetic counselors, fertility clinics, and "dating/matchmaking services":

> In one example, a Virtual Progeny assessment can be offered to a customer in connection with a matchmaking service, for example, through a single company or a co-marketing or partnership relationship. A user of a matchmaking service can order an assessment of Virtual Progeny described herein to determine the probability that an offspring resulting from the potential match between the user and a candidate partner will express one or more traits described herein. The user can then use this information to aid in evaluating the candidate partner for a potential match. The matchmaking service can be an on-line service, such as Shaadi .com, eHarmony.com and Match.com.

I wonder what genetic counseling might look like at eHarmony.com and if anyone would actually opt to learn about the urination pattern of Virtual Progeny before going on a first date. I also wonder how colossally vague, only partly genetic qualities like "mental acuity" or "social intelligence" or "learning/memory" or "maternal/paternal behavior" could be even theoretically estimated from a couple of spit samples in the first place, but with those qualities, not to mention "breast size/shape" or "grooming behavior," we are far outside even the most expansive conception of health.

In the *New Scientist* article, the company's founders appeared to disagree about the algorithm's future application. Anne Morriss was "adamant that the firm doesn't intend to use the system for non-medical purposes." Her cofounder was more equivocal: "Silver says that what the patent is used for in future will be a business decision. He also says that owning the patent means the firm can prevent others from using the technology in unintended ways." The disagreement illustrates a fundamental opposition in the rhetoric of biotechnology: reassurance and excitement. The reassurance is attached to more restricted, familiar applications (such as enhancing health). The excitement is attached to more expansive, novel applications (such as enhancing people). The persuasion in service of biotech tends to split the difference, presenting new applications as exciting yet safe, revolutionary but reliable.

It would be easy to pivot here to a dark warning. To suggest that the future will play out along the lines sketched out in *Remaking Eden*, with health shading into enhancement, then leading to unrecognizable posthuman descendants. But it's one thing to turn a book idea into a real-world company and quite another to extrapolate to a genetically engineered future. That extrapolation assumes that progress will, more or less, chug along, which assumes any number of systemic and political stabilities we cannot assume. So I focus on what we're saying in the present, on the fables that create the future.

In her book *Feminist, Queer, Crip*, Alison Kafer considers, from a disability rights perspective, the complex traffic between future visions and present realities. "As critics of utopian thinking have long argued," she writes, "the futures we imagine reveal the biases of the present; it seems entirely possible that imagining different futures and temporalities might help us see, and do, the present differently." To me, Kafer's statement has two useful implications. First, learning to imagine the people who are actually here, in the present, may help us to bring about better futures; and second, those we fail to imagine today will be either absent from the future or alive only on its margins. Kafer's book title is loaded: *Feminist, Queer, Crip* is both a condensed bio and a charged reclamation of insults. But the title also questions the categories by which people are divided and suggests that those categorical boundaries are not absolute, that one can inhabit multiple intersecting categories, and that the way we name them matters.

Kafer does not reject medical intervention; she rejects the understanding of disability as purely medical. She argues, along with most in disability studies, that disability occurs at the intersection of the body and the world. She also argues that "[t]he category of 'disabled' can only be understood in relation to 'able-bodied' or 'able-minded,' a binary in which each term forms the borders of the other." In the "relational model" of disability she advocates, "disability is experienced in and through relationships; it does not occur in isolation." She rejects the idea that "any future that includes disability can only be a future to avoid":

> In this framework, a future with disability is a future no one wants, and the figure of the disabled person, especially the disabled fetus or child, becomes the symbol of this undesired future. As James Watson—a geneticist involved in the discovery of DNA and the development of the Human Genome Project—puts it, "We already accept that most couples don't want a Down child. You would have to be crazy to say you wanted one, because that child has no future."

Watson's view is common, of course. It is underwritten by the belief that disability's meaning is both negative and fixed, a function of the objective body and not of the fine mesh of assumptions and curb cuts and laws in which the body exists. Against this view, Kafer asserts another one not yet attained: "In imagining more accessible futures, I am yearning for an elsewhere—and, perhaps, an 'elsewhen'—in which disability is understood otherwise: as political, as valuable, as integral."

*   *   *

Since its inception, GenePeeks has marketed its company mission with a true story about family. Anne Morriss, the cofounder and CEO, has a son with medium-chain acyl-CoA dehydrogenase (MCAD) deficiency—a potentially fatal disorder, which went undetected when Morriss used a sperm donor to conceive. Morriss's research led her to Lee Silver and then to the company's founding. That narrative is now told in a YouTube video embedded on the company's home page. (The condition is treatable with diet, and Morriss's son is doing well.)

The video opens with an establishing shot of Morriss's palatial house. Inside, she narrates her story to an off-camera interviewer, focusing on the moment of diagnosis—an indelible moment for nearly all parents, and one often botched by doctors. This was true for Morriss: a few days after she got home, she got a phone call and was asked if her son was still alive. In a surreal, reality-TV touch, Morriss reenacts the moment: in footage spliced into the interview, we see her picking up the phone, listening intently, looking shocked and concerned. There's a close-up of the receiver lying ominously on the desk, as if she has just run to the nursery. The background music is somber, the lighting muted.

But then, as in an antidepressant commercial, the mood changes. We see Morriss and Silver walking together, chatting easily but seriously, and as the video moves into the open floor plan of GenePeeks's headquarters, the music brightens. We see animated young professionals, talking around a conference table. In a commercial, we'd see the same turn from despair to happiness, but where the commercial would begin with a woman staring longingly out at the rain, pivot with the announcement of a brand-named drug, and end with the same woman laughing nonanxiously around a campfire with her family, here the change is signaled by images of a thriving workplace. The closing shots return us to Morriss's home, where we see her son playing joyfully among iridescent soap bubbles on an expansive back deck. Inside, Morriss delivers the clincher: "Genetic diseases have become preventable conditions, and that's our mission as a company. We want to help you protect the health of your future family."

It is a complex, slickly produced narrative, but it would be wrong to deny Morriss's evident love for her son or her sincerity in wanting to help others. The reality TV–style reenactment does not diminish the actual shock of finding out that your child's life is at risk, just as the formulaic cheesiness of ads for antidepressants does not mean depression is unreal. In fact, it's the grain of emotional truth that makes the whole thing work: the most effective persuasion has roots in something authentic, in shared values charged with feeling and made artfully legible, and the video does just that, using Morriss's true-life story to associate a proprietary algorithm with images of family, work, and home—of a good American life.

And yet the video undercuts its own marketing message: the vision of family happiness *includes a child with a preventable condition*. On the one hand, that child's diagnosis is shown as a shocking moment, an avoidable tragedy. On the other, the child is an irreplaceable part of a happy family. His presence in the video introduces uncertainty: he stands for both loved family member and feared outcome. But more than that, he is the essential beginning of the story. Because of her son, Morriss delves into the science, meets Lee Silver, and cofounds a company. In this reading, the happiness and purpose she discovers in founding GenePeeks is a direct consequence of her son's arrival. Even as the video narrative draws on the familiar trope of the sick child as tragedy, it draws on a second narrative in which a child transforms a parent's life, spurring growth and purpose.

* * *

It's an odd experience to see a child portrayed, in an expensively produced video on the home page of a genetic testing company, as both a tragedy and a gift. Odder yet, when the portrayal stars the child's parent. But in that collision of meanings is a lesson about the impact of biotechnology: it both exposes and entangles two competing visions of the family. One suggests that family happiness depends on health, which is guaranteed by technology; another suggests that happiness depends on accepting a child for himself, whoever he is. Only the first, though, fits with selling a product whose essential purpose is selective.

There are several problems with using a negative portrait of disability to sell biotechnology: it skates over the complexities of actual families, implies that disabilities are necessarily obstacles to happiness, and lays responsibility on the consumer—not the system of social supports—to guarantee the best chance of flourishing. Most of all, it contributes to the misconceptions that people with disabilities face daily. But a less obvious problem is the broader assumption that technology and disability are opposed, that the main purpose of technology is to either fix disability or prevent it. This assumption, in turn, obscures the fact that we *all* depend on technologies, high and low, to flourish in the world. At the same time, it frames people with disabilities as targets or passive recipients of technology, not authorities on their own lives or sources of understanding. On this point, here's Sara Hendren, a designer, writer, and disability theorist:

> One of the things I say a lot when I lecture is that people with disabilities who are using assistive technologies every day are our richest resource of wisdom about the cyborg-self, about how we integrate technologies into our lives. They've been doing it in significant ways already, every day, for a long time. But because we have these notions of assistive technologies as medicalized, and a kind of a medical-tragedy story we tell ourselves about these people, we have ignored them as a knowledge resource.

Hendren has a son with Down syndrome, whose arrival changed the way she sees the world. "After my son was born," she says, "I just looked around. My antennae were already picking up visual culture because that's my training. I started to keep this running taxonomy of the built environment, typefaces, his glasses, leg braces, and adaptive toys. This was a brand new world, and I was captivated by it." She questions the term "assistive technology": as she points out, all technology is assistive. Similarly, she embraces high technology and low, technologies for health and technologies for living. Both/and, not either/or:

> I want a technology that, yes, preserves independence in old age when it's wanted. I want a technology that preserves health as long as it can, and augments health when it's called for. But what I often call for, too, is technologies for interdependence. The enduring human always needs assistance. The goal is not, in other words, elimination of assistance or elimination of exchanges with one to another. The goal is thriving communities over the whole life cycle.

I don't worry about, nor am I excited by, a distant future of human upgrades, of wings, flippers, and superpowered brains. I do worry about technologies both founded on and sold with a distorted portrait of disability, an imagined future diversity that neglects the real range of brains and bodies in the present. If biotechnology is to contribute to a just society, then a nuanced, representative understanding of disability, disease, and human frailty must be part of the conversation. But our deliberations about biotechnology are hampered by a triple invisibility: of the technology that stands to affect disability, of nuanced thinking about disability's meaning, and of people with disabilities themselves.

2   The Germline

The practice of Eugenics has already obtained a considerable hold on popular estimation, and is steadily acquiring the status of a practical question, and not that of a mere vision in Utopia.

—Francis Galton, *Essays in Eugenics*

Francis Galton, the Victorian polymath, innovator in forensic science and statistics, and inventor of the word "eugenics," took it for granted that improved intellect was good, that it would be better for society if the more intelligent outbred the less, and that therefore persuasion was called for. In the brief introduction to his 1909 *Essays in Eugenics*, Galton dwelled less on science than persuasion, writing that the key to "Eugenic reform" was "Popular Opinion, whose dictates … become so thoroughly assimilated that they seem to be original and individual to those who are guided by them." In the collection's second chapter ("Eugenics: Its Definition, Scope, and Aims"), Galton set forth a sequence of persuasive goals:

> *Firstly* [eugenics] must be made familiar as an academic question, until its exact importance has been understood and accepted as a fact; *Secondly* it must be recognized as a subject whose practical development deserves serious consideration; and *Thirdly* it must be introduced into the national conscience, like a new religion.

Deploying the metaphor of an alphabet, Galton argued that "civic worth" was obviously inheritable, and that citizens could be divided into classes A through X, from "criminals, semi-criminals, loafers and some others," up through "a large mass of mediocrity," to those who "found great industries, establish vast undertakings, increase the wealth of multitudes and amass large fortunes for themselves"; that those at one end of the scale were, in every sense, worth more and should be encouraged to breed, while the "loafers" should be discouraged. He further believed that since heredity was destiny, charity was pointless. (In an eccentric touch, the beginning of the alphabet signified the lowest ranks. A more conventional example came from biologist Edward M. East, who wrote, "[I]f in the future the proportion of people of Grades A and B increases, the nation will prosper; while if the proportion of people of Grades D and E increases, the nation will decay.")

As persuasion, the alphabet metaphor was inspired. It was simple, memorable, linear. It expressed a hierarchy of value, with the wealthy, white, and intellectually able at the top. Because its elements were discrete—letters, unlike colors, do not blend into one another—the metaphor reinforced the idea that people could be classified into objectively discrete groups; because an alphabet is linear, it lent itself to a hierarchy of both individuals and races, obscuring geographical or cultural variation. But it also suggested the classifier's power: the truth was not what people believed about themselves, it was what the expert could observe. Grades imply the grader's authority. (In very real terms, the line between *normal* and *abnormal* remains the line between *pass* and *fail*.)

I think of Galton as the Montaigne of eugenics: a founding figure, singular, omnivorous, and encyclopedic. His *Essays in Eugenics* are essays in more than name: they are speculative and exploratory, and they consider the meaning of a good life. But Galton was uninterested in the way an individual made sense of experience; he focused on that which was fixed, inheritable, and quantifiable. How to live mattered less than which lives mattered in the first place. In Galton's reverse alphabet of achievement, Ws and Xs contributed to progress; Ms and Ns, "the great mass of mediocrity," did not; and Bs and Cs detracted. The differences were fixed by heredity.

Galton's hierarchy was centered on intellect but charged with class, race, and morality. Historian Nathaniel Comfort writes that Galton "first floated the idea for engineering society in 1865, in an article for the general-interest magazine *Macmillan's*":

> He fantasized about an entire race of idealized upper-class English intellectuals. ... To get there, he invoked an image of animal husbandry that became a standard trope of American eugenicists in the twentieth century: "If a twentieth part of the cost and pains were spent in measures for the improvement of the human race that is spent on the improvement of the breed of horses and cattle, what a galaxy of genius might we not create!"

No one seriously suggests breeding human populations for improvement today, but the value we place on intellect—and the corresponding devaluation of intellectual disability—persists. That orientation is reflected in our visions of the future: under the rubric of health, the elimination of intellectual disability is seen as desirable, and in more extravagant visions, "a galaxy of genius" is the end point. Intellect defines a destination, the means to reach it, and the ground of value.

*   *   *

It is difficult to know where we are in the story, which is why we keep trying to tell it. That's the story where *Homo sapiens* becomes *Homo faber* or *Homo deus* or *H. sapiens 2.0*, where we engineer the germline, altering the species in inheritable ways, directing our own evolution. A century and a half ago, Galton thought this project was "acquiring the status of a practical question," but he had human breeding in mind, not direct manipulation. But with the advent of the CRISPR-Cas9 system—a tool for making precise and permanent changes to the DNA of living cells—*what was once science fiction, we are told, is now a reality*. It's a familiar claim, but its tired language bears a spark of truth: that comprehending the present, no less than the future, requires an act of narrative imagination. That something like science fiction is necessary when the future arrives ahead of schedule.

CRISPR-Cas9 is a human technology adapted from bacteria and archaea, single-celled animals which evolved the system to defend against attacking viruses. When a bacterium equipped with CRISPR survives a viral attack, it "remembers" the viral sequence, storing it between repetitive lengths of DNA—*c*lustered *r*egularly *i*nterspaced *s*hort *p*alindromic *r*epeats, abbreviated CRISPR—which are spaced in the bacterium's genome like beads on a string. Future invasions of the same virus are then met with a cas (for "CRISPR-associated") protein, which targets the remembered sequence and snips the virus's DNA, rendering it harmless. This system has been reengineered to edit genomes in living cells: a "guide RNA" can be designed to identify *any* short sequence in the DNA of a living cell, and a cas protein (usually cas9) to cut the DNA of that cell in a precise location. Many of the applications are uncontroversial: engineering experimental mice in a few weeks, for example, that would otherwise take months or years to breed. Cancer therapy is another possibility: CRISPR could be used to disable tumor cells. But CRISPR can also be used, in principle, to edit the DNA resident in sperm, egg, or embryo—the germline—and these changes would be inherited.

CRISPR is sometimes described as a "word processor," able to "cut" and "paste" lengths of DNA at selected locations, a metaphor which subtly oversells CRISPR's accuracy. Though it's faster, more precise, and easier to use than previous genome-editing tools, its edits, unlike a word processor's, don't always occur as intended. Sometimes a cut occurs in the wrong place ("off-target effects"), or it doesn't occur at all. Two recent studies also found that CRISPR-edited cells may be more likely to develop cancer. That said, I am less interested in CRISPR per se than the way we narrate it to ourselves and the assumptions that inform the story. CRISPR will be improved, perfected, or superseded, but what matters is the way we think about people and the form the story takes. Technologies change, but persuasion springs eternal.

The standard story is that a new story is about to begin. Jennifer Doudna, the codeveloper of the CRISPR-Cas9 system, writes in *A Crack in Creation* (cowritten by Samuel Sternberg) that "[w]e're standing on the cusp of a new era, one in which we will have primary authority over life's genetic makeup and all its vibrant and varied outputs." This rhetoric predates CRISPR. In *Life at the Speed of Light,* scientist and entrepreneur Craig Venter writes that we are at the "dawn of the digital age of biology," and that "humankind is about to enter a new phase of evolution." In his book *Regenesis*, a far-reaching exploration of new genomic technologies, Harvard biologist George Church frames the turn in explicitly narrative terms: the genome is "the greatest story ever told," and we're about to start rewriting it. *Regenesis*: In the beginning was the Word. Church's title, like Doudna's *A Crack in Creation*, revises religious language for secular ends, but the very act of revising old myths undermines the announcement of a new age. We are negotiating between present and past to create the future, and the claim that a new age is dawning is the hoariest one of all.

But perhaps the claim is true, and we are about to cross a line, to begin directing our own evolution. Perhaps we will look back and say that the line was crossed long ago, that the actual decision to modify the first human germline was trivial, that technology's momentum could not have been stopped. Or perhaps there is no line, only a region of uncertainty where we draw lines to comfort ourselves. Perhaps the claim of a new chapter in history only reflects the economy of attention in which the claim is made, an economy in which dramatic claims are required to sell books and drive clicks. If this is so, to accept the claim is to add to the hype, when clarity demands subtraction.

* * *

Late in *A Crack in Creation*, Jennifer Doudna recounts a meeting with a venture capitalist eager to fund the first CRISPR baby. Doudna depicts herself as uncomfortable, the investor as sketchy and too eager. It's a brief morality tale embedded in a narrative of discovery: though most of the book is devoted to her work developing and optimizing the CRISPR-Cas9 system, her goal is also to open up discussion of the technology's implications. In the scene, the venture capitalist is essentially a personified abstraction: she represents Incautious Greed, a foil to Doudna's Reasonable Reflection.

I prefer *A Crack in Creation* to most other scientist-authored books about the genetic future. Doudna's book shows someone happiest at the bench; interested in improving human health and reducing suffering; and uninterested in provocation, grandstanding, or transhumanist fantasy. The book's tone is even and measured, though there are flashes of something else: desire for credit, anger at the "betrayal" of unnamed others. Doudna is also able to admit uncertainty, even fear. She narrates a dream about Hitler: "He had a pig face and I could only see him from behind and he was taking notes and he said, 'I want to understand the uses and implications of this amazing technology.'" She mentions disability rights, of which many public scientists seem completely unaware.

Doudna's book is a good faith effort to spur conversation, and she writes that the scientist's role is "to introduce and demystify … technical accomplishments so the public can understand their implications and decide how to use them." But Doudna's approach is persuasive, not just informative—note that she says *how*, and not *whether*—and in making the case for CRISPR's power, Doudna offers long lists of the problems it can solve. The result is a textbook case of the medical model of disability, in which a wide range of conditions are lumped together as curable pathologies:

> [A] quick survey of the published scientific literature reveals a growing list of diseases for which potential genetic cures have been developed with CRISPR: achondroplasia (dwarfism), chronic granulomatous disease, Alzheimer's disease, congenital hearing loss, amyotrophic lateral sclerosis (ALS), high cholesterol, diabetes, Tay-Sachs, skin disorders, fragile X syndrome, and even infertility. In virtually all cases where a particular mutation or defective DNA sequence can be

linked to a pathology, CRISPR can in principle reverse the mutation or replace the damaged gene with a healthy sequence.

Powerful technologies translate categories into human facts; the conditions classed as abnormal are the ones we test for and may one day edit. These categories are more powerful than any slur, and their expansion is driven by persuasion: the more clinical targets CRISPR can access, the more powerful it seems. Because of CRISPR's enormous commercial potential, this tactic is fueled by the profit motive as well. The market forces that drive the technology's development also drive our understanding of disease.

Though Doudna describes her position on germline editing as "evolving," she comes down, with caveats, in favor of its eventual use. Central to her argument is the invocation of those of suffer. Parents of children with "devastating" genetic disorders are key, if fleeting, characters; she also quotes Charles Sabine, a man with Huntington's disease—a fatal, late-onset, single-gene disease—as arguing that there is "no moral issue at all" in using inheritable gene editing for a cure. She asks, "Who are we to tell him otherwise?"

For me, this is at once a good question and a terrible question. It's a good question for me because it reminds me that each of us peers at the world of treatment and cure through a different lens, and that you can't automatically export the wisdom of one frame to another; you have to begin by listening. To have a daughter with Down syndrome sheds light on these issues, but my experience, my politics, and my conclusions are different from those of other parents in the same situation—let alone people facing Huntington's in their families. But using a patient's voice as a trump card doesn't open up debate; it shuts it down. As a tactic, it parallels other doubtful strategies in the argument for new biotechnologies. Tagging opponents as "emotional" or "Luddite" disqualifies them, making discussion pointless. Arguing that human gene editing is "inevitable," as Doudna and many others do, radically restricts the conversation: if the technology is inevitable, then the only question is how to apply it. Tactics like these tend to normalize the technology, to make it seem routine. It would be wiser to remember its strangeness.

That Huntington's entails suffering, that a cure is desirable, is completely uncontroversial among people with the disease, their loved ones, and biomedical researchers. In her book *Mapping Fate*, the writer and historian Alice Wexler offers a history of the disease as a family member; her mother died from the disease, and her sister Nancy led the team that discovered first the genetic marker that made a test possible and then the gene itself.

It's a remarkable book, a mix of science and literature, showing a single disease in multiple dimensions: personal, familial, social, historical, genetic. She shows that the experience of the disease varies from family to family but shows that the disease, despite variations, is inexorable, and its slow onset, a gradual but perceptible loss of abilities, is itself a form of suffering. Alice does not herself have Huntington's, but her account sounds less like waiting for the shoe to drop than watching it fall in slow motion:

> Living at risk undermines confidence, for there is no way of separating the ordinary difficulties and setbacks of life from the early symptoms of the illness. It is not like any other physical illness, where consciousness can at least continue in the knowledge that one is still oneself, despite severe pain and physical limitation. Huntington's means a loss of identity.

In this, and in its late but certain onset, Huntington's is utterly unlike deafness, achondroplasia, or Down syndrome. This is precisely why Doudna's use of the disease is questionable: it anchors a broader argument for gene-editing in which a wide range of conditions are classed as abnormal, and abnormality leads to suffering, where lists of conditions potentially curable by CRISPR include everything from Tay-Sachs to deafness. There's an unresolved tension between using the worst cases (like Huntington's disease) to boost the case for using CRISPR, and using long lists of milder conditions to widen our sense of CRISPR's scope.

In a famous TED Talk, the late activist Stella Young addressed the idea that disability necessarily entails suffering. Having critiqued the practice of "inspiration porn," the images of people "overcoming" their disabilities, she continues:

> [L]ife as a disabled person is actually somewhat difficult. We do overcome some things. But the things that we're overcoming are not the things that you think they are. They are not things to do with our bodies. I use the term "disabled people" quite deliberately, because I subscribe to what's called the social model of disability, which tells us that we are more disabled by the society that we live in than by our bodies and our diagnoses.
>
> So I have lived in this body a long time. I'm quite fond of it. It does the things that I need it to do, and I've learned to use it to the best of its capacity just as you have, and that's the thing about those kids in those [inspirational] pictures as well. They're not doing anything out of the ordinary. They are just using their bodies to the best of their capacity. So is it really fair to objectify them in the way that we do, to share those images? People, when they say, "You're an inspiration," they mean it as a compliment. And I know why it happens. It's because of the lie,

it's because we've been sold this lie that disability makes you exceptional. And it honestly doesn't.

It is critical to listen to patient voices and to take them into account. But what constitutes suffering, what causes it, the relation between suffering and ordinary experience, and the place of technology in its remediation are and always will be open questions. For this reason, care is required in using terms like *disease, disability, abnormality,* and *suffering.* All too often, though, these concepts eddy and merge. "The symptoms of hyperargininemia," Doudna writes, "are awful; they include progressively increasing spasms, epilepsy, and severe mental retardation." Discussing her unease with research on primates, she writes,

> Gene editing is also being exploited to target genes implicated in neural disorders, taking advantage of the fact that monkey models are uniquely suited for the study of human behavioral and cognitive abnormalities. Although on one level, I feel uneasy about using monkeys in this way, I am also sensitive to the intense need to develop treatments and cures for human disease to alleviate human suffering.

"Behavioral and cognitive abnormalities" is an extremely wide category. Equating these with "disease" is problematic, with "suffering" exponentially more so. To think or behave differently, to be "abnormal," does not necessarily entail suffering. But for Doudna, "behavioral and cognitive abnormalities" are so awful that they justify overcoming a reluctance to use primates for research. If you peel back the claims of health, you often find ideas about intellect. This is especially true when future people are under discussion.

*   *   *

In December of 2015, *The Guardian* reported that the London Sperm Bank had rejected a donor for having dyslexia. The bank had also published a 2010 pamphlet with a long list of disqualifying "neurological diseases," including dyslexia, autism, attention deficit hyperactivity disorder (ADHD), and other conditions. Vanessa Smith—described as a "quality manager at the JD Healthcare Group," the bank's parent organization—defended the bank. Backpedaling without budging an inch, she said that the pamphlet had been withdrawn and policies would be reviewed. Still, little seems likely to change. According to Smith, "We are looking for someone who is medically clear of infectious diseases and genetic issues that may possibly be passed on to any resulting child." She also claimed, "We definitely don't work in eugenics." But to shape future children, based on a policy that describes human variation as disease, is eugenic by definition.

Avoiding hepatitis C is not like avoiding dyslexia. *Infectious diseases* is specific, but *genetic issues* less so; the euphemistic vagueness of "issues" enables a double emphasis on pathology and genes, even when neither is fully appropriate. Autism is not a disease. Neither is dyslexia. Neither is unambiguously genetic, in the way that Tay-Sachs or Down syndrome is, and cerebral palsy can occur without genetic influence at all. But since these conditions *may* have a significant genetic component, they're on the list. This is the one-drop rule for the new millennium: any *hint* of a disorder that may or may not be genetic is, in this scenario, sufficient to disqualify its bearer.

Specifics imply caring. To group radically dissimilar conditions as "genetic issues" suggests little about those conditions and a great deal about the idea of normality to which they are opposed. Tellingly, however, the pamphlet is obsessively specific about Different Brains, even to the point of redundancy: forbidden are ADD (attention deficit disorder) *and* ADHD, autism *and* Asperger's, "mental retardation" *and* Down syndrome. (Men with Down syndrome are thought to be sterile.) That emphasis is typical. Though the pitch for new biotechnologies is made on the basis of health, with the promise of curing or preventing genetic disease, an unquestioned valuation of intellect—and a corresponding devaluation of people with intellectual disabilities—underpins the enterprise.

Babies just arrived are tangible, radiant, squalling; they resist our projections from the beginning. But future children are abstract, nonexistent, easy to shape. No violence is done, it seems, when one possibility is chosen over another. In practice, this means that the business of future children trades in wishes, in providing a sharper image of the child already half-imagined. Intellect is central to the image. In an article on American sperm banks, *Washington Post* reporter Ariana Eunjung Cha wrote, "Fertility companies freely admit that specimens from attractive donors go fast, but it's intelligence that drives the pricing: Many companies charge more for donors with a graduate degree." But the companies do more than list credentials: they offer thumbnail profiles, character sketches. Little fictions, lending ghostly bodies to hope. Cha wrote,

> One cryobank, Family Creations, which has offices in Los Angeles, Atlanta, Austin and other large cities, notes that a 23-year-old egg donor "excels in calligraphy, singing, modeling, metal art sculpting, painting, drawing, shading and clay sculpting." A 29-year-old donor "excels in softball, tennis, writing and dancing."

The Seattle Sperm Bank categorizes its donors into three popular categories: "top athletes," "physicians, dentists and medical residents," and "musicians."

And the Fairfax Cryobank in Northern Virginia, one of the nation's largest, typically stocks sperm from about 500 carefully vetted donors whose profiles read like overeager suitors on a dating site: Donor No. 4499 "enjoys swimming, fencing and reading and writing poetry." Donor No. 4963 "is an easygoing man with a quick wit." Donor No. 4345 has "well-developed pectorals and arm muscles."

With the exception of the musclebound Donor No. 4345, the key feature of every one of these donors is intelligence. But that intelligence isn't cited directly: it's made imaginable in terms of achievement. An implicit chain of causation is identified, beginning with good genes and ending with someone successful, creative, or both, whether it be a doctor, a lawyer, or a boring dilettante for whom "swimming, fencing, and reading and writing poetry" are equivalent activities. What is sold is a future adult who has it all, both physical and mental superiority—which is a very different dream than a child who is healthy. What's being sold, to those who can afford it, is a child who is *better*.

How much do these representations matter? The fortunes of people with dyslexia do not rise or fall on the basis of a single pamphlet from the London Sperm Bank; the diversity of human embodiment is not catastrophically threatened by a sperm bank's marketing. Nor does a single book, patent application, or op-ed determine the future of disability. But neither are these representations neutral. Each is a tiny, glittering chip in a vast mosaic of common sense, a market-driven portrait of whom we welcome.

*   *   *

Near the conclusion of *Regenesis: How Synthetic Biology Will Reinvent Nature and Ourselves* (coauthored by Ed Regis), Harvard biologist George Church suggests what sorts of people we may welcome to the engineered future:

> We may embrace much greater human diversity, not merely ancestry but vast spectra of personality, age, and intellectual capacities (e.g. an intentional increase in high-functioning autistics, bipolars, and ADHDs). This may require very specialized and highly trained parenting—well beyond the current random assignment of child to parent.

I am all for embracing a greater spectrum of intellectual capacities, but in this case Church seems to mean only those from the far right end of the bell curve, those smart enough to compensate in some way for what would otherwise be considered defective. The disabled will be admitted, as long as they are able enough. High-functioning disabled people are worthy of an "intentional increase"; others, presumably, are not. Of course, the "high-functioning autistics, bipolars, and ADHDs" might understand their identities in other ways. They might also resent being born for a purpose. The place of the "low-functioning," or the possibility of a parallel, intentional decrease, is not discussed. Nor is the political arrangement in which a nondisabled "we" embraces select disabled people and decides which populations to intentionally increase.

Church's speculations sound novel but are rooted in a long tradition: evaluating people only insofar as they *can* contribute and attributing failure to the people themselves, not the society that restricts their welcome. By suggesting that birth parents would not be competent to raise the children, that these (engineered?) progeny would require special technical care, Church ignores the actual history in which institutional horrors have often been perpetrated by experts, and parents have been a driving force for the rights of their children—not to mention the more practical ideal, in which professionals *help* parents to raise their children. But in Church's model, function outweighs relationship.

Perhaps Church's suggestion is only a provocation. A side dish, a condiment, beside the banquet of the future. Even so, it's consistent with the economic focus of his book, whose metaphors are saturated with the history and values of industrial capitalism. Recounting the iGem competition, where college students build synthetic organisms—"genetically engineered machines"—to win prizes, Church celebrates industrialization: the winning entry "recapitulated" the superiority of "assembly-line mass production" to the "prior system of individual craftsmen working by themselves." Elsewhere, he describes cells as "tiny, obedient molecular factories." Synthetic biologists sound like a cross between industrialists and colonial explorers, "civilizing, taming, and domesticating the basic processes of life." Discussing the manufacture of new biomaterials, Church deploys a more digital metaphor, saying that "a maximal genome" would resemble "the vast online shopping world":

> We call this collection of nearly all enzymes *E. pluri,* as in *E. coli* meets *E pluribus unum.* "From many states to one nation" becomes, in the world of genome engineering, "From many enzyme genes to one integrated, customizable, all-purpose biofactory."

In this metaphor, the country itself is like a factory, so it's unsurprising that citizens would be valued by function. At the same time, the use of industrial metaphors, like the use of religious ones, suggests that the new age of information is haunted by the old.

We cannot separate biotechnology from the way we talk about it. If biotech is more than the tool itself, but the set of practices and understandings that make it possible—what historian of science Hallam Stevens calls a sociotechnical system—then, in the outer reaches of that system, are the metaphors by which a particular biotech application is sold. Reading those metaphors closely can illuminate the assumptions likely to drive the technology's use. Comparing genomes to factories celebrates the synthesis and control of life; comparing them to "an online shopping world" illuminates the extent to which life has become commercialized; citing *e pluribus unum*, the slogan on our currency, extends those concerns to American identity; suggesting a conditional welcome for "high-functioning" people with cognitive differences shows us which human differences are valued, and devalued, in the present.

But close reading can also help us see that no metaphor is inevitable, and that therefore different ones are possible. Researching this book, trying to wear down the vast mountain of my scientific ignorance, I found *The Way Life Works*—an illustrated introduction to biology, a collaboration between a molecular biologist and an illustrator. That book also invoked the phrase *e pluribus unum*, but in a very different way: the phrase subtitled a chapter on *Community*, whose point was that all life is interconnected and interdependent. Life, according to the authors, is a wondrous, evolved system in dynamic balance, with its self-adjusting feedback loops, its balance of competition and cooperation, its ingenious adaptations and endless forms driven by evolution. The illustrations were quirky, colorful, and approachable. Life was described as information, but it was figured as a grand river, not a code. *The* Amazon, not Amazon, "the vast online shopping world."

I'm a longtime Oregonian: faced with these competing understandings of *e pluribus unum,* I'm going to take the river over the world of online shopping. But the lives of people with disabilities exist beyond these tidy metaphorical visions, and both visions leave open the question of where people with functional differences fit. Will they be part of the many, part of the one? If nature is a harmonious system, then some can seem dissonant. If life is interdependence, a complex miracle of interlaced adaptations, then some can seem maladapted, broken, abnormal. Indeed, in their emphasis on perfect function, the two visions of *e pluribus unum* are not so far apart: one features a biofactory and the other an evolved, untrammeled nature, but both idealize a whole that depends on the proper working of its individual parts.

* * *

To predict a future of human engineering is to begin to imagine it, and one standard line on that future, the staple of news reports and books by scientists, is that the future holds "promise" and "peril." This frame is inherently reassuring: it suggests that there are only two options, both of which can be envisioned. Having two simple alternatives, even when one is frightening, is preferable to a seething, unreadable complexity.

That binary, however, is insufficient to its subject. It normalizes a radical, species-altering event as a familiar kind of choice, in which risks and benefits can be calmly assessed: two roads diverge, as it were, and all we have to do is take the right one. But gene editing is more like terraforming, changing the landscape, the idea of roads, the people who walk on the roads. If undertaken at large scale, it would alter us in ways both direct and indirect: some by being engineered, some because they undertook the engineering, and all of us by living in a world where the power was embraced. We would be defined by our stances toward it, our choices to embrace, refuse, accommodate, resist. It is not merely disease that is at stake, but identity, and in more ways than we can calculate, human gene editing would rewrite the meaning of pronouns—*I, we, you*—from the inside.

For me, the real question is this: Promise and peril for whom? The risks and benefits are not exclusive; what matters is *who* takes on the risk, *who* benefits. The idea that "we" can all walk down the path marked Promise, if we are informed and wise enough, suggests that we are all in this together, and owing to the drastic inequalities both within our society and worldwide, we are not. The fierce patent battles over CRISPR suggest that it will need to be monetized, and the promises made on its behalf suggest that profit, more than actual need, will decide which applications get to market. Promise and peril are unlikely to be equally distributed. People with disabilities, who are disproportionately poor, are unlikely to be targeted as consumers. They are, however, likely to be cited as justifications for the technology's use.

For many with disabilities, the problem is the way "promise" and "peril" are defined in the first place: the terms are at best insufficient, at worst hostile. For people who embrace disability with pride, as a social, cultural, and political identity, the promise of eliminating genetic disease can *be* the peril. A present invisibility can lead to a future disappearance. In an essay entitled "Please Don't Edit Me Out," for example, disability activist Rebecca Cokley raised concerns about gene editing, suggesting what it is to be on the wrong side of "progress":

> Stories about genetic editing typically focus on "progress" and "remediation," but they often ignore the voice of one key group: the people whose genes would be edited.
>
> That's my voice. I have achondroplasia, the most common form of dwarfism, which has affected my family for three generations. I'm also a woman and a mother—the people most likely to be affected by human genetic editing.
>
> I remember clearly when John Wasmuth discovered fibroblast growth factor receptor 3 in 1994. He was searching for the Down syndrome gene and found us. I remember my mother's horrified reaction when she heard the news. And I remember watching other adult little people react in fear while average-height parents cheered it as "progress."

Against a world obsessed with abnormality, many try to reframe the idea of normalcy; against the erasure of ordinary experience, many try to reinscribe it, to insist on everyday experiences in which inconvenience and discrimination loom larger than genes. It is an asymmetrical opposition, not least because the rhetoric of biotechnology deals in projected future people, as if a white light were shining through a diagnostic image, while people with disabilities are three-dimensional individuals living in a specific time and place. The late bioethicist Adrienne Asch, who was blind, summed up the disconnect:

> People with disabilities are thinking about a traffic jam, a disagreement with a friend, which movie to attend, or which team will win the World Series—not just about their diagnosis. Having a disability can intrude into a person's consciousness if events bring it to the fore: if 2 lift-equipped buses in a row fail to stop for a man using a wheelchair; if the theater ticket agent insults a patron with Down syndrome by refusing to take money for her ticket; if a hearing-impaired person misses a train connection because he did not know that a track change had been announced.

The flood of personal accounts of disability online can be seen, broadly, as part of a larger argument about the meaning of experience and as an assertion that disability as lived experience is different from disability as categorized from without. Some accounts bear witness to prejudice: to the impact of being devalued, considered defective or broken, pitied or scorned, and to the exhaustion of having to explain, justify, and defend. Some explore the uncertain territory where disability and disease overlap. Most reject the simple equation of "abnormality" and "suffering," the assumption of tragedy; many reclaim a difference as a positive good, a different way of being in the world. That approach is neither sentimental nor intrinsically opposed to technology, as Rachel Kolb demonstrates in her essay about music, cochlear implants, and deafness:

> When I got a cochlear implant seven years ago, after being profoundly deaf for my entire life, hearing friends and acquaintances started asking me the same few questions: Had I heard music yet? Did I like it? What did it sound like?

Kolb insists on the social context of deafness. She highlights the curiosity of others, the attitudes of the nondisabled. A dinner companion tells her it's "sad" that she can't hear the music in the restaurant. This casual pity occasions self-questioning at the time, but it also occasions the essay itself, which is framed as a years-late reply. In doing so, Kolb splices two times together, a remembered question and a present answer:

> *Sad.* This is how some hearing people reacted to my imagined lifetime without music. Did it mean that some part of my existence was unalterably sad, too? I resisted this response. My life was already beautiful and rich without music, just different. And even if listening to music did not yet feel like a core part of my identity, I could be curious.

For Kolb, the technology of cochlear implants is transformative but not redemptive. The device's effect is sometimes "harsh"; she finds relief in being able to turn it off. It opens up new dimensions of music to her, but it does not rescue her from a tragic defect; it helps her understand the richness of the experience she already had and leads her, paradoxically, to an understanding of music that transcends hearing:

> Not only does music ingrain itself in our bodies in ways beyond simply the auditory, it also becomes more remarkable once it does.
> "Can you hear the music?" Even though I now can, I think this question misses the point. Music is also wonderfully and inescapably visual, physical, tactile—and, in these ways, it weaves its rhythms through our lives. I now think a far richer question might be: "What does music *feel like* to you?"

It's common to frame an essay as the-thing-I-could-not-say-then: something troubles the writer, sets a chain of thinking in motion, and the essay is the result. Kolb's reply to a question asked long ago illuminates the joys and social complexities of her way of being in the world; in her nuanced take on the cochlear implant, she shows the complications and rewards of technology in actual use. Technology is neither rejected nor fetishized. It does not rescue her from difference; it becomes part of her ongoing adaptation to the world. Kolb's account is *open*, in the way a patchwork list of "defects," "mutations," or "abnormalities" is not.

*   *   *

Reading books like *Regenesis* or *A Crack in Creation*, I was struck to see that disability was not just the subject of metaphor; it was also a source. Disability provides the language by which nature, and the technology that would alter it, is understood. The authors think about disability, but they also think *with* disability, using it to express a generic dismissal.

It's common, for example, to describe nature as "blind"—a disability-based metaphor used to characterize random evolution as inferior to the directed variety. George Church asserts that "the course of synthetic genomics will be under our own deliberation and control instead of being directed by the blind and opportunistic processes of natural selection." Jennifer Doudna describes a case in which a man recovers spontaneously from a genetic disease: "[I]magine that the human genome is a large piece of software...the blind programmer in this instance was nature itself."

In these figures, nature is personified as disabled. The point is that nature is not as *good* as we are: we *aren't* blind. Interestingly, "blind" is a nested metaphor, meaning here "without conscious intention." The metaphors are interlinked, like cogs in a machine of misunderstanding: a sensory disability stands for an intellectual one, which in turn establishes a hierarchy of ability. We, the engineers, are the sighted and rational ones, while nature is blind and disabled, without intention, the one who makes mistakes. In a revealing variation on this theme, biotech applications are personified as intelligent: describing "gene drive," which helps engineered organisms outcompete their counterparts in the wild, George Church writes that it "appears to outsmart natural selection." The metaphor is not only about ability; it is about power. Or judgments about ability are always about power.

Obscured by these figures are people with disabilities themselves, as human beings with interests, differences, personalities, and stories. Our rhetoric is paradoxical, at once invoking and erasing them. Like the metaphors we live by, people with intellectual disabilities are invisible—which is to say, present but unnoticed. They are most likely to be affected by the new technologies affecting the human, but least likely to be represented. This is one consequence of an intellect-segregated world, where intellectuals who produce both the technologies and the opinions spend most of their time in universities and the intellectually disabled live in group homes or with their parents. In the absence of overlap, it is easier for the second group to become an abstraction.

That invisibility is apparent in the historical record. The words and ideas of eugenicists are abundant, but the words of their targets are rare. The old textbooks, the issues of *Eugenical News*, the bulletins of the Eugenics Record Office, letters to wealthy donors, and the proceedings of the Race Betterment Foundation are archived, studied, dissected, a river whose headwaters begin with Galton, its words pooling in archives and books and Internet sites devoted to disability history. But there is a river of silence beside the river of record, the unspoken words of a population much talked about but little heard, a population either invisible or obscured by misunderstanding.

*   *   *

Francis Galton dreamed of a "galaxy of genius," a theme echoed by his American successors. David Starr Jordan, the first president of Stanford University and a leader in American eugenics, wrote that "[an] aristocracy of brains is the final purpose of democracy"; J. H. Kellogg, an early scion of the cereal manufacturing family, described the goal of eugenics as "an aristocracy, a group of men and women who are willing to keep themselves unspotted from the world, a nucleus from which in time may develop a new and better human race."

It is impossible to trace simple parallels between the era of mainline eugenics and today: we are not repeating the past. But neither have we escaped it, and the dream of an "aristocracy of brains" via directed evolution is still alive in the writings of futurists. In *Remaking Eden*, Lee Silver speculates that the future holds "a special group of mental beings":

> Although these beings can trace their ancestry back directly to *homo sapiens*, they are as different from humans as humans are from the primitive worms with tiny brains that first crawled along the earth's surface.

Ramez Naam's pro-enhancement *More than Human* ends similarly:

> Long after we are gone, after there no longer live any beings with our DNA, our distant descendants will still look back. They'll look at this moment in our history

and marvel. *To think that such primitive creatures as* Homo Sapiens *could give birth to whole new kingdoms of life! They could not have understood what they were creating! Yet we're lucky that they had such a strong urge to change and grow—that is why we're here.*

It is Galton's "galaxy of genius," made literal: an intellectual aristocracy, displaced into the distant future, even if they sound suspiciously like the Advanced Beings who appeared regularly on the syndicated *Star Trek* reruns of my youth. The difference is that on *Star Trek*, the writers were pulling for the humans, who were plucky and flawed and passionate and loyal, while the Advanced Beings were typically rational but cold. They had lost something in their self-perfection. For Silver and Naam, though, the future beings seem both superior and more interesting. We are *primitive* before them, we are like *worms* with *tiny brains*. We are, in comparison, intellectually disabled.

3 At the Fair

[T]he people of Oregon have safeguarded the breeding of fancy horses and cattle. They should at least do as much for their people.

—State Senator Sylvester Farrell, Republican of Oregon, 1917

In the depths of my hard drive is a photograph of Laura, at thirteen, posing next to an alpaca at the Benton County Fair. They are dressed up as Elsa and Anna, the sisters from *Frozen*. Magno, the alpaca, is draped with pink fabric, and his head sports butterscotch-colored pigtails made from a braided hank of yarn. Laura's hair is sprayed silver, and she is wearing a long blue dress, elbow-length white gloves, and pink New Balance sneakers.

As a then-member of the Lucky Longnecks 4-H Club, Laura had spent a year working with Magno, who lived on a farm outside town, cropping grass and tending his buried species memory of the Andes. Laura had learned to walk Mags forward and backward and in a circle. She had brushed his soft coat and learned the names of a couple of alpaca diseases, and one day she even gave him vitamin shots with a veterinary syringe. Laura has since moved on to keeping and showing rabbits, and Mags has moved on to the great Peruvian pasture in the sky, but we remember him fondly. He was old and gentle, with an unruly mop of hair and spectacular buckteeth, like a cross between an early Beatle and a camel.

The *Frozen* costumes were part of the Large Animal Costume Contest, a ritual at the intersection of drag show and animal husbandry. For it, the Lucky Longnecks led their camelids in an impromptu parade around the fairgrounds in the blazing heat. Laura had hoped to sing one of the *Frozen* numbers for the judges and was disappointed when this turned out not to be an option. Among all the fears and hopes and uncertainties and predictions I entertained after Laura arrived, I can safely say that no scenario involved an aging male alpaca dressed up to resemble a fictional Disney princess at the Benton County Fair. The costumes were Laura's idea, though Theresa had done most of the sewing.

We like 4-H for the same reason we like the Special Olympics Unified Basketball Team: it's a place where children of different abilities can be together on equal, if asymmetrical, terms. But in 2017, as I researched this book and Laura researched rabbit diseases so that she could answer the judge's questions during Showmanship, the history of American eugenics cast a long shadow across the fair. It was sweltering, pushing 109 degrees, and as the sweating kids plumped their rabbits on individual carpet squares, waiting for judgment—Best in Breed, Best in Show—the whole enterprise seemed an exercise in Bunny Eugenics, a separation of the fit from the unfit, a careful examination of Bunny Physiognomy to determine the local pinnacles of the breed.

The directed evolution of people is inseparable from the directed evolution of animals: from Dolly to gene-edited mice to CRISPR-engineered miniature pet pigs, animals are the models we use to work out the details of the future. But historically, to invoke animals was to invoke intellectual disability: nonhuman animals defined the lower boundary of a human hierarchy, which "feeble-minded" people approached or crossed—a racialized hierarchy, with whiteness at the top and nonwhite races at the bottom. That complex set of prejudices is distilled in the phrase *Mongolian idiocy*, and it underlies Down's racial system, but the loose association of race, disability, and nonhuman animals is still endemic online.

For all these reasons, Laura's presence at the 2017 Benton County Fair—a full member of the Claws and Paws of the Round Table 4-H Club (not to be confused with the Benton Rabbiteers, the Critters of the Valley, or the Happy Hoppers), a high school junior with a Facebook account and a bedroom wall banded with blue ribbons and Star Wars stickers at the high-tide line of her outstretched fingers—is a marker of genuine progress. In the early decades of the twentieth century, at the height of American eugenics, a child classified as "feeble-minded" would have been present, if at all, as an idea of what to avoid. For a time, in fact, it was the people, as well as animals, competing for ribbons at the fair.

The eugenic impulse is always accompanied by persuasion, and one of the stranger persuasive forms in the United States began before World War I, as contests for Better Babies and Fitter Families. "Purebred farm animals were an obvious and widely used analogy for better babies," write Dorothy Nelkin and M. Susan Lindee, "and the prize ribbons and medals were like the standard awards of the stock show." In the Better Babies Contests, babies were dressed in identical white togas and examined competitively for adherence to eugenic norms. In the Fitter Families Contests, entire families underwent individual physical examinations and answered extensive questionnaires with the goal of identifying the model families, the ones free from visible or inherited defect. The Fitter Families Contests always featured an intelligence test. In mainline eugenic thinking, "feeble-mindedness" was the central defect of interest. Though allegedly detectable via IQ scores, "feeble-mindedness" had a far-ranging, almost mystical power: it was the source of crime, poverty, and sexual deviance, and for this reason it mattered very much which families should have children and which should not.

The Better Babies and Fitter Families Contests first sprang up locally, at state fairs in the Midwest. After the war, they spread quickly, growing in popularity; in the 1920s, they came to be sponsored by the American Eugenics Society (AES), an influential group dedicated to education. The AES used the contests both to spread the gospel of better human breeding and to collect data from the questionnaires. The Society also sponsored visual displays which explicitly linked human breeding, animal breeding, and the costs of "feeble-mindedness." A poster from the 1929 Kansas Free Fair reads,

HOW LONG

ARE WE AMERICANS TO

BE SO CAREFUL FOR THE

PEDIGREE OF OUR PIGS

AND CHICKENS AND

CATTLE,—AND THEN

LEAVE THE ANCESTRY

OF OUR CHILDREN

TO CHANCE, OR TO "BLIND" SENTIMENT?

The block letters are handwritten. Despite the implied ethos of perfection, the lines are not quite parallel. The significance of the quotation marks around "blind" is unclear, but the parallels with contemporary rhetoric—in which rational control is opposed to chance, sentiment, and "blind" nature—are unmistakable. If human improvement is on stage, disability-based metaphors are usually skulking in the wings.

Another display from the same fair, bearing the headline SOME PEOPLE ARE BORN TO BE A BURDEN ON THE REST, offered carefully printed placards beneath single light bulbs:

> This light flashes every 15 seconds
>   Every 15 seconds $100 of your money goes for the care of persons with bad heredity such as the insane, feeble-minded, criminals, and other defectives.

And:

> This light flashes every 48 seconds
>   Every 48 seconds a person is born in the United States who will never grow up beyond that stage of a normal 8 year old boy or girl.

Taken together, the crude placards and the contests attempted persuasion on multiple fronts. However crude, that persuasion—though it claimed the language of statistics and reason, the implied authority of the objective—appealed to the emotions with the techniques of art and story. A blinking light, to denote urgency. A Fitter Families Contest, to elevate a particular eugenic ideal into a social ritual, as a hybrid of competition and public display. In the age of smartphones, these may seem quaint, but not as much has changed as we might like to think. The eugenics posters, with their light bulbs, are multimedia displays. The Better Babies and Fitter Families Contests, in which people were poked, prodded, and above all measured according to norms, turned persuasion into competition, fun, and participatory ritual. All have their far more sophisticated equivalents today, and now as then, our norms are quantitatively defined but haunted by qualitative ideals.

The point of the Fitter Families Contests, and of the crude, lighted displays, was to get "better" families to breed more, and "worse" families to breed less. Taken together, they defined an imaginative world in which class, race, morality, national identity, and inheritance were manifested in, and depended on, the family. That imaginative world was elaborated with the eugenic family studies: spurious genealogical research, in which entire family trees were profiled, showing the consequences of a single, ill-fated match. The Kallikaks, the Jukes, and other pseudonymous clans supposedly proved that bad matches led to every social ill, to prostitution, alcoholism, "pauperism," and crime, and that all were linked to "feeble-mindedness." At no other time in the nation's history have people with intellectual disabilities been so demonized. In 1927, the notorious Supreme Court decision in *Buck v. Bell*—as a result of which Virginia asylum resident Carrie Buck was sterilized against her will—opened the door to legalized forcible sterilization. Tens of thousands of Americans, men and women both, were sterilized afterwards; the practice continued into the 1970s, with sterilizations occurring in the California prison system as recently as 2010. In a bleak illustration of our interconnectedness, "feeble-mindedness," a historically pliable category, expanded far beyond any reasonable definition. Predictably, women of color and poor whites were disproportionately affected.

Whether you see differences or continuities between the heyday of American eugenics in the 1920s and today depends on where you look. The differences aren't really in dispute: The pseudoscientific association of intellectual disability and criminality, the crudest forms of racism, the hereditarian focus, and the belief that complex, ill-defined behaviors like "pauperism" could be inherited like eye color, have faded from mainstream science. Many of these have been discredited *by* science. And most of the rhetoric attached to new biotechnology emphasizes health and individual choice, not a responsibility to the nation's future.

But the continuities endure. It's still common to invoke disability as a cost to society; racism is thriving; the ideals of health, beauty, strength, and most of all intellect are largely the same; and, in the commercial space, the arguments are advanced by images of ideal families and children. For the thousands of men and women sterilized in the decades after World War II, it cannot have been much consolation that mainstream scientists had rejected the errors of their pseudoscientific forerunners. For many on the margins, the shift in scientific thought had little to do with their lives. Their problem was a political one, and it would take activism to begin to change it.

* * *

In the early eugenicists' visions of directed evolution, better animals formed the template for a vision of better humans. Mary T. Watts, one of the founding organizers of the Fitter Families Contests, wrote that "while the stock judges are testing the Holsteins, Jerseys, and Whitefaces in the stock pavilion, we are judging the Joneses, Smiths, and the Johnsons." Both the bovine and human Whitefaces were populations to improve and protect.

In eugenic thinking, that analogy was structural, not decorative. American eugenics had its institutional beginning within the American Breeders' Association, whose Eugenics Section included most of the figures who would go on to national prominence.

But if purebred animals were used to illuminate an idealized human norm, a generic idea of the animal was used to downgrade the humans considered defective. The hierarchy found its expression in a kind of metaphor: "animalization," according to historian Gerald O'Brien, a literary form of stigma. O'Brien notes that eugenics was "[b]uilt on the concept of animal breeding" and as such "by its very nature carried with it an undercurrent of animalization." As O'Brien notes, this habit extends well beyond intellectual disability and was and is used to "dehumaniz[e] undesirable community members."

In the rhetoric of eugenic leaders of the 1920s, as O'Brien points out, that "undercurrent" rises to the surface. Leon Whitney—the executive secretary of the AES, which sponsored the Fitter Family Contests—compared "feeble-minded citizens" to adopted bear cubs: cute when young, a menace when grown. Charles Davenport, head of the Eugenics Record Office, "frequently spoke of the 'unfit' segment of the population as being more like animals than like humans." Davenport wrote, "If we are to build up in America a society worthy of the species *man* then we must take such steps as will prevent the increase or even the perpetuation of animalistic strains."

> In other works he contended that "in some way or other society must end these animalistic bloodlines or they will end society," and that "by the elimination of the worst matings of the animalistic strains and by the union of sense and sentiment in many others a more uniform innate capacity in our people may be achieved."

But to focus on statements like these in isolation is to miss the complex, on-the-ground realities of Fitter Families and Better Babies Contests. The women who actually ran the contests were true believers in the eugenic program, and their work was an opportunity to spread the gospel. But they were also interested in public health: maternal health, infant mortality, and nutrition. Alexandra Minna Stern, tracing the history of Better Babies Contests in Indiana, writes about physician Ada Schweitzer, who "implored Hoosiers to adhere to the state's marriage laws and spoke out consistently in favor of the state's sterilization restrictions." At the same time, Schweitzer was also committed "to the gospels of private hygiene, pure milk, vaccination programs, and clean air and sunshine." Personal health and national destiny were joined, a strange combination expressed in the Indiana Child Creed, part of an "advice manual…distributed free of charge…to every mother who registered her newborn with the state." While the creed asserted a child's right to health—"to be loved; to have its individuality respected…to be protected from disease, from evil influences and evil persons; and to have a fair chance"—it also asserted, "Every child has the inalienable right to be born free from disease, free from deformity and with pure blood in its veins and arteries."

Schweitzer's public health work was devoted, year-round, and significant: in Indiana, infant mortality was cut almost in half. But in the end, Schweitzer believed that nature trumped nurture: "You cannot make a silk purse out of a sow's ear, neither can we make a citizen out of an idiot or any person who is not well born." Belonging hinged on heredity. The "idiot," the person "not well born," was not a pitiable case to be cared for, but an active threat to be excluded. It is not surprising that Schweitzer uses an animal metaphor, *a sow's ear*, to express disenfranchisement. To have an animalistic nature was to be irredeemable, to be barred from the community of the species.

Like the organizers, the participants had complex motives rooted in ideas of American identity and progress. Historian Laura L. Lovett writes that "the fitter family contests…should not be interpreted as a piece of mass culture deftly produced for passive consumption." They were persuasive, but their audience had its own reasons for participating. They may have wanted a free medical exam, or been interested in genealogy, or welcomed the chance to proclaim their rural roots. Families, according to Lovett, eagerly touted their pioneer roots in the required questionnaire. Their nostalgia, Lovett argues, should be seen against a backdrop of increasing industrialization, and the contests both embraced and resisted progress: "[T]hese contests fused nostalgia for the farm family with a modernist promise of scientific control." The contests constituted a negotiation between familiar and unfamiliar, old and new. But that negotiation, as the contest's full name suggests, began with home and family: *Fitter Families for Future Firesides*. The contests were about the American hearth and who would sit beside it.

At the same time, they offered coherence in the face of change, a place in time. The contests were forward- and backward-looking at once, celebrating a traditional idea of family in the lens of science and modernity, appealing to the durable fantasy of progress without disruption. New and old, religion and science, the past and future of family, community and country, were reconciled in a competition.

The bronze medal given to Fitter Family Contest winners bore a quote from the Psalms: *Yea, for I have a goodly heritage.* That quote spun *heritage* into a triple helix of significance: religious, national, and biological. The winners were white, Christian, American, deemed fit by science: members of the genetic elect.

At the Kansas Free Fair, in the Better Babies Contests, the heritage had a classical twist: contestants wore identical white flannel togas. The grading was extremely gentle, with miniscule fractions of a point subtracted so no one would feel bad. Parents were anxious anyway, and some earnestly wrote the judges about their children's defects and failures, asking how they might improve their scores in the following year.

* * *

I surf the photographs online. The babies in togas, the Fittest Family posed and smiling. *Yea, I have a goodly heritage.* I have no roots in that Midwest world. In 1929 my mother was two years old in Tokyo. My father would be born a year later in New York City. He grew up in Queens. Some of his mother's family, left in Poland, were slaughtered by the Nazis. In the late 1930s the letters stopped coming. Other Jews sailed for America and were sent back to be killed. The immigration restrictions that sentenced them to death shared a taproot with Fitter Families and Better Babies Contests and with the sterilizations of women like Carrie Buck and Fannie Lou Hamer and thousands of others whose names have been forgotten or lost.

There's a famous line drawing of a tree labeled *Eugenics*, the logo for the Second International Congress of Eugenics, held at the American Museum of Natural History in 1921. A stout, Keebler-elf kind of tree, its snaking roots labeled with fields of study: *genetics* most prominently, but also *mental testing, anthropometry, religion, medicine, surgery, statistics, psychiatry, law, politics, biography*. The font is Tolkienesque, the image homely and ridiculous, and if it were not for the human misery directly traceable to the idea, and the motto emblazoned in the open sky visible between the roots and branches—*Eugenics is the self direction of human evolution*—it might seem comical, dated. But to see only the comedy is to believe that the idea is safely in the past and to miss the ambition behind the image. Eugenics was the sum of human knowledge: an entire university, redrawn as a single poison tree.

The image is ubiquitous, familiar, but I had to see it many times before it occurred to me that the roots are labeled, but the branches are not. They are just regular tree branches with what look like oak leaves, though they could be labeled *castration, hysterectomy, institutionalization, immigration restriction*, everything that grew from those toxic roots. The image is, like so much of the rhetoric of eugenics, self-referential. A carefully shaped image of a living thing, arguing that living things should be shaped. A natural thing, turned into a message. It is also a transformation of other iconic trees: family tree, tree of life. It revises past images as it advocates the revision of the race.

Looking closer, I also noticed that not every human endeavor is represented. *Art* is absent. So are *folklore, poetry, rhetoric*. The illustration, of course, *is* a persuasive form of art, a laboriously detailed message, a rhetoric both written and visual, an instance of metaphor. Its absences show what it truly is, just as the children absent from the Better Babies Contests— "feeble-minded," black, Jewish, Asian—showed what they really were, what drove them. The rural pioneer ideal to which the good midwesterners harkened was defined by the people it left out and displaced.

These annotations are meant to fill the absences: to show the American sunlight falling on the tree, the soil where it could sprout and grow, the water that sustained it. To show that the tree is still growing.

*   *   *

In *Regenesis*, George Church draws an explicit connection between engineering humans and engineering animals. Church argues that "engineering recapitulates evolution": "we stand at the door of manipulating genomes in a way that reflects the progress of evolutionary history: starting with the simplest organisms and ending, most portentously, by being able to alter our own genetic makeup." From cells to people, or, in Church's phrase, "from bioplastics to H. Sapiens 2.0." One project already underway, nearer to us than to single cells, is "de-extinction": the effort to "bring back" extinct species using remnant DNA, including the passenger pigeon and the woolly mammoth, and to reintroduce them in the wild. Of many scientists worldwide working on de-extinction, Church is the most famous, and the one most willing to push boundaries: where others focus on de-extinction simply as a conservation tool, Church famously suggested that along with the woolly mammoth, Neanderthals could be brought back, and that the first Neanderthal might be gestated in an "adventurous female human."

De-extinction has been much discussed in print, but the most complete case for the project is made at the website of Revive & Restore, a nonprofit dedicated to "genomic conservation," whose stated goal is "to enhance biodiversity through new techniques of genetic rescue for endangered and extinct species." Revive & Restore is the project of husband-and-wife team Stewart Brand and Ryan Phelan. Brand is the founder of the *Whole Earth Catalog*; Phelan is an environmentalist and entrepreneur. Revive & Restore touts, sponsors, and helps to coordinate several hoped-for restorations, including the heath hen, the passenger pigeon, and the woolly mammoth.

As with any radically new biotech application, advocates of de-extinction face a set of persuasive challenges: explaining and normalizing a complex technology; answering, avoiding, or finessing ethical objections; and making a radically new approach to nature seem emotionally "right." To study the Revive & Restore site is to see that computers have expanded the possibilities not only for genetic manipulation, but for persuasion. The online medium is complex and interactive: words, images, and video combine and reinforce each other, and appeals to reason and emotion are interwoven. The persuasive approach, though its details are specific to de-extinction, holds for other biotech applications as well. The application in question is presented as an ideal route to an ideal outcome, and technical and conceptual complications tend to be soft-pedaled or erased. Most of all, persuasion begins with the nouns, with the way the technology is named.

The power of "de-extinction" resides in the prefix: it negates a negative. "De-extinction"—neutral, scientific sounding—frames the project as extinction's opposite, assuming the questionable premise that extinction can, in fact, be reversed. It's a tidy but misleading label. De-extincted creatures will not be "brought back" so much as engineered. The mammoths or pigeons in question will be created species that physically resemble old, evolved ones.

On the Revive & Restore site, the ordinary language of de-extinction is already persuasive. Feel-good words—*conservation, comeback, restoration, rescue*—are typically paired with technical words: *genomic conservation, genetic rescue*. These phrases, like the technology itself, splice old and new together. But they are also part of a larger rhetoric of information, which itself expresses a world caught between digital and analog. *Server farm, digital breadcrumbs, DNA library*: we are still in bodies, and we still need the physical to imagine information, scaling it up through metaphor, making it a little more tangible. We imagine DNA as words, sentences, chapters, and books. We harvest data, imagining farms. We wander through the virtual, imagining a trail of crumbs behind us, as if we were children in a folktale.

At the Revive & Restore site, idealized language is anchored by images of the ideal to be achieved, by artist's conceptions of the passenger pigeons and mammoths to come: digital reproductions of paintings, realistic, frozen, a taxidermy of the future. A composed still life of lost species, with the passenger pigeon front and center; a single pigeon depicted against blue sky, wings outspread; a woolly mammoth looming toward the viewer, its curving tusks dominating the foreground, the curvature of the earth barely visible behind it. In context, digital reproductions of paintings are more than a little ironic. Though their lush realism evokes another century, a time when many extinct species were alive and abundant, the images function differently in practice. Digitally reproduced, distributed on the Internet, viewed on screens, the images are computer-dependent reproductions used to advocate for computer-dependent reproductions. They are, no less than a de-extincted mammoth, a blend of old and new.

The realism, in other words, is compensatory. It reveals the uneasy tension between old and new brought on by powerful biotech: the more artifice the project requires, the more its advocates retail images of naturalness and purity. Like the photos of perfect babies that fill the ads for prenatal tests, the oil paintings of majestic mammoths are idealized images of the natural, presented as the result of advanced technology. The fear that our technologies will change nature is met with images *of* nature. From this perspective, the background of each image is as significant as its central focus: the meadow around the serenely pregnant woman, in the ad for a new prenatal test; the tundra behind the woolly mammoth, pristine as the curving, painted background in a natural history museum diorama.

This pattern holds for other biotech applications. In 2012, for example, the company then known as the Gene Security Network—a pioneer in noninvasive prenatal testing, or NIPT—renamed itself as Natera. The banner of the revamped home page featured a sun rising over a field of flowers. When you looked closer, the sun was actually a clump of cells. In the sky was the company's vaguely military-sounding motto: *Conceive. Deliver.* The name change was accompanied by an explanation, an instance of self-conscious persuasion:

> The name Natera is drawn from the terms: natal, nature, and earth. Our new name better reflects our mission to help couples around the world manage pregnancies and reduce the risk of genetic disease.

For me, at least, Natera's words and imagery had the opposite of their intended effect: rather than softening science and technology with images of nature, they present an image of nature in the grip of technology. The word *Natera*, in splicing the word *nature* with other words, produces something unnatural: an etymological GMO, as it were, as sentimental and clinical as a blastocyst rising like a sun.

From *Gene Security Network* to *Natera* may be the most extreme name change in the history of American corporate governance: what had sounded like a shadow organization in one of the Bourne movies now sounded like a shampoo. To me, the change demonstrates a sharpened focus on NIPT's consumers, a move to a more "feminine" appeal. (That shift from a masculine-sounding to a feminine-sounding name, from protection to nurture, was matched by Sequenom, whose SEQureDx test became MaterniT21 PLUS.) Recent images are more expertly natural: flowers tremble in a light wind; a woman walks, in slow motion and soft focus, through a field of ripened wheat. As with ads for antidepressants, images of nature compensate for their opposite. Images of what is "natural" are conjured against the artificiality of the intervention.

Changes like these point to our anxieties about an altered nature, but they also point to the market economy in which the changes take place. In the broadest sense, the idea of pristine nature is itself generated by an industrial society; something is only "pristine" in comparison with its opposite. In the narrow sense, every large-scale technological application, profit or nonprofit, needs money to continue. The appeal of restored species, as Stewart Brand noted in a 2014 interview, is crucial to attracting donors. With refreshing frankness, he explained that mammoths are easier to fund than mice: "As architects say, form follows funding. The animals that will draw avid supporters who have avid amounts of money will probably be the first ones that get dealt with." The projects chosen, in other words, are the ones most likely to be approved and funded—which means that questions of persuasion are integral, not incidental, to the project. But in its web-driven emphasis on the visual, de-extinction's persuasive appeals shift our gaze from the unsung and rapidly disappearing to the easily anthropo-morphized and mediagenic. Mammoths and passenger pigeons lend them-selves to web pages; bees don't.

Brand is both reflecting on persuasion and trying to persuade. But in his scenario, the animals are themselves persuasive: by inspiring others, they will help to protect the ecosystems they inhabit:

> Such animals can also serve as icons, flagship species inspiring the protection of a whole region.

And:

> Conservationists are learning the benefits of building hope and building on hope. Species brought back from extinction will be beacons of hope.

It's a hopeful narrative about hope, in which an animal-induced feeling leads to a concrete policy result:

> The return of the marvelous marsupial wolf called the thylacine (or Tasmanian tiger), extinct since 1936, would ensure better protection for its old habitat.

The word helps us imagine the tiger's return, the return helps us imag-ine the value of nature, and habitat is thus preserved. The creature will be persuasive in the future, just as words about the creature are persuasive in the present. In this sequence, the animal itself becomes a message. Behind this logic is an understanding of life as a form of information. That idea underpins the de-extinction project, in which genomes—ancient, or newly created—are information to manipulate and recombine. But the projected animal itself becomes a form of information, a word which, if spoken to life, might change the world.

Inspiration is easy, but habitat protection is hard. M. R. O'Connor's *Resurrection Science*, exploring different forms of what Brand calls "genetic rescue," illuminates the intractable political, economic, and ecological complications involved in attempts to save even a single existing species. To write that the very return of an animal, even the Tasmanian tiger, would *ensure* habitat protection is optimistic in the extreme. Our world abounds with inspiring animals, of which stunning photographs exist. The animals are still going extinct, and their habitats are still vanishing. O'Connor writes, "I found the individuals working on de-extinction projects to be brilliant, and a few downright inspiring. But not many have shown how we will put resurrected animals back into the world at a time when humans can barely coexist with extant species."

The optimism behind Brand's approach reaches its zenith in one of the main selling points for reviving the mammoth: that it will help combat climate change. In this formulation, the mammoth becomes more than a message; it becomes a tool. Brand writes, "[The mammoths'] return to the north would bring back carbon-fixing grass and reduce greenhouse-gas-releasing tundra." By grazing (and fertilizing), it will foster the rapid return of grassland; by compacting the earth, it will help to keep methane locked in permafrost. In this paradigm of inspiration, the mammoth—an engineered creature—will itself become an engineer, an agent of planetary change. It will be made in the image of its creators.

* * *

To believe that synthetic mammoths may one day walk the earth is reasonable. To believe that vast herds of synthetic mammoths will not only successfully populate the Siberian wastes, but also reengineer the ecosystem, thus helping lower the Earth's temperature, requires something like an optimism cascade. Everything has to go the way it is supposed to, and the law of unintended consequences has to take an extended holiday. This is why the project makes me think of Rube Goldberg, the inventor and cartoonist. Which, in turn, makes me think of my dad.

He was an engineer—a really good one, according to the coworkers I talked to after his death. He also liked Goldberg's cartoons, which he probably first saw as a kid growing up in Queens. I wish I could remember more, but I don't. Still, I have a thread with something live on the end, if not a clear memory of the labyrinth, and at the other end was my father, the heating, piping, and air-conditioning engineer. He designed cooling systems for nuclear power plants, he was at ease with complication, he trusted numbers and logic, and he liked the cartoons, which depicted elaborate contraptions to perform trivial tasks: a back scratcher, a device to hand you soap in the shower, plausible on the page and implausible in life. Maybe he liked the art: he had taken a drawing class as a young man. Maybe they gave shape to his understanding that life tended to go haywire. He lived by facts and data but did not think they were everything. Our ordinary conflicts could be framed, crudely, as a conflict between engineering and art, the practical sciences and the impractical humanities, but that frame would be as false as the division itself. The weather between us was variable, very rarely stormy, but there was bedrock beneath the weather.

As he got older, my father read more and more. Fiction, mostly. He never got all the way to poetry; I never read much fiction. We recommended books to each other but didn't read them. I imagine a vanishing point at which we converge, where I am patiently waiting, or he is. At that imaginary point, we both let go of our belief that facts and rationality are enough. I see now that it haunted us both like the tattered remnant of a shroud, some scrap of proverbial wisdom we clung to in the faith that the whole text might be recoverable. If we'd had time, we might have come to see that belief for what it was: not something to discard or live by exclusively, but a key sentence in a larger text we were writing together.

I suppose if my memory were upgraded, I might be able to access the precise day and time in Bergenfield, New Jersey, when we laughed about a specific cartoon, and if social media had been a thing in the early 1970s, then I could just scroll back through the timeline to when I tagged Dad with an image of the Rube Goldberg machine and he typed Haha or pressed Like, but not really, because if he were alive today I doubt he'd be caught dead near Facebook, which is to his eternal credit. But maybe I'm wrong about that too.

Looking the cartoons up now on Google Images, I'm struck by an odd fact: the cartoons almost always include animals, usually anthropomorphized, sometimes imaginary. A partial list: sentient bear, zozzle hound, elephant, pet dragon, seal, Albanian Ook, Giant Umpha Bird, Russian Dancing Bug, homesick goldfish. The inventions only work if the animals behave exactly as expected, at the exact right time. Here are the instructions, for example, for the personal fan, which depicts a man pushing a long-handled wheelbarrow:

> Take hold of handles (A) of wheelbarrow (B) and start walking—pulley (C) turns kicking arrangement (D) which annoys bear (E)—bear suspects doll (F) and eats it, pulling string (G) which starts mechanical bird (H) saying, "Do you love me?"—love-bird (I) keeps shaking head "yes," causing fan (J) to move back and forth making nice breeze blow right in your face.

I think of the cartoons now, reading about the plan to mitigate climate change by breeding mammoths. It seems a Rube Goldberg machine on a planetary scale. Find mammoth. Dig up mammoth. Unfreeze tissue. Rescue partial genome, splice into elephant genome, insert into enucleated elephant egg, shock elephant/mammoth egg to life, implant egg in newly invented artificial uterus, gestate mammoth, deliver mammoth, repeat on massive scale, transport to Siberia, train in herd behavior, keep healthy, and presto! All this, of course, assumes an absence of political conflict, whether local or international. There are nearly a million people in the vicinity of "Pleistocene Park," the land in Siberia where, it is hoped, mammoths will one day roam. The park is the brainchild of Sergey Zimov and is now managed by his son Nikita, who was profiled by Ross Andersen in the *Atlantic*. There are already other Arctic animals there: horses, the odd musk ox. When Andersen asked Nikita Zimov how he kept local hunters from shooting the animals already there, Zimov replied that connections made the difference: he had spoken at the funeral of the local Mafia leader.

I'd love more than anything to talk about all this with my father now, just as I'd love for him to meet Ellie and Laura, and that fact alone suggests the power of the wish to resurrect, to reclaim what is lost. I get the wish to bring back mammoths; I get the powerful desire for something to go right, for once, in a world that is rapidly warming and losing dozens of species every day. But I remember the Rube Goldberg machine.

*   *   *

In her book *How to Clone a Mammoth: The Science of De-extinction*, Beth Shapiro, a biologist associated with Revive & Restore, makes no claims about climate. She also rejects Brand's atonement argument (we should bring back the mammoth because we made it go extinct), and unlike George Church, she does not appear to see de-extinction as part of a larger program of directing evolution in people. About the very idea of "bringing back" or "restoring" lost species, she is blunt:

> I will argue that the present focus on bringing back particular species—whether that means mammoths, dodos, passenger pigeons, or anything else—is misguided.…Extinct species are gone forever. We will never bring something back that is 100 percent identical—physiologically, genetically, and behaviorally identical—to a species that is no longer alive.

For Shapiro, de-extinction (and related projects) are about ecosystems, not species. She criticizes "our partiality toward charismatic megafauna," which are "selected for de-extinction based on their public appeal." At times, she seems impatient:

> … regardless of how excited I feel about our latest results, the most common question I am asked about them is, "Does this mean that we can clone a mammoth?"
> Always the mammoth.

Early in her book, Shapiro writes that she wants "to separate the science of de-extinction from the science fiction of de-extinction." In this framing, science provides certainty; "science fiction," a literary genre, is equivalent to hype and exaggeration. This is a common frame, in which "hype" is blamed on nonscientific others. (The common targets are the Media, Activists, and the Science-Ignorant Public.) In her account of the TEDx conference on de-extinction, for example, Shapiro blames the media for sensationalizing the project and for failing to detect her note of caution. The scientist is level-headed and rational, the media and public ignorant and excitable.

This approach is problematic for several reasons. First is the fact that Revive & Restore, as we've seen, is the central clearinghouse for de-extinction hype. Second, as anthropologist Sophia Roosth suggests in *Synthetic: How Life Got Made,* the project itself shades into science fiction. What is proposed is the invention of new life-forms and the engineering of ecosystems, and its advocates trade in images of planetary engineering, time travel, and raising animals from the dead. "Pleistocene Park" is a play on "Jurassic Park"; George Church suggests that "adventure tourists" will one day visit the site; Sergey Zimov envisions "bullet trains on elevated tracks" to avoid disturbing the animals. Church also describes the place as a "Siberian eden," a Biblical allusion. The textual paradise presages the engineered one.

So the science cannot be boiled off from the fiction, any more than the persuasion can be separated from the project. In fact, Shapiro offers her own flight of Siberian fancy. In imagining the point of it all, she describes a pristine landscape:

> … the perfect arctic scene, where mammoth (or mammoth-like) families graze the steppe tundra, sharing the frozen landscape with herds of bison, horses, and reindeer—a landscape in which mammoths are free to roam, rut, and reproduce without the need of human intervention and without fear of re-extinction.

Tellingly, the star of this scene toggles between "mammoth" and "mammoth-like": like Shapiro's book, which is titled *How to Clone a Mammoth* but says that neither cloning nor mammoths are likely, the scene builds excitement with a vision of the real thing while undercutting our notion of what is itself real. It's a strange idyll, an imagined wilderness "without the need of human intervention," in which the keystone species are synthetic. This tension is more than an aspect of rhetoric; Roosth argues that it's central to the project, where "[s]ynthetic nature…masquerades as a wild primeval landscape" and "life-forms not yet made become props for creatures already gone." *Masquerades, props, fantasies*: the project is as literary as it is scientific, a technically adept act of imagination.

That imagination is enacted on two levels. The animals themselves are creations, which need to be imagined before they are created. But prior to those future acts of flesh-and-blood imagination are acts of imaginative persuasion, fictions constructed to help bring the real thing about. The artist's rendition precedes the animal, and the stories precede the herds. But if you look closely, the stories are themselves chimeras, like elephants rewired to have extra fat and hair. On the outside, they look like stories about preservation; on the inside, they're stories about engineering.

The Revive & Restore site has gained a degree of nuance over the years. Its claims, though still bold, are less bald. It acknowledges, though in fine print, some of the objections and questions surrounding de-extinction. And yet its most prominent message, embodied by inspiring renderings of mammoths, by graphics that frame living and extinct species in the same panorama of time, remains the same. The website itself, in other words, has evolved: an engineered animal, recoded in response to a changing environment.

*   *   *

Fitter families, better babies, de-extinction, metaphor, rhetoric, NIPT, Laura, bunnies, mammoths, Oregon, home: I'm building my own Rube Goldberg device, writing the way a synthetic biologist might, if the synthetic biologist were a nonlinear thinker with a daughter with Down syndrome. Perhaps this is only serious play, making with a question attached, a nonstandard thing built from the Registry of Standard Intellectual Parts, a chimera made of words. If this is the case, so be it: I pursue the connections to see if there are any, and I try to assemble disparate things into a working picture. To find coherence, a stay against confusion.

So as I think about mammoths becoming more common, I think about Down syndrome becoming more rare. Live births of children with Down syndrome are currently about a third less than what would be expected, and with the increased uptake of NIPT, this trend is likely to accelerate. I don't think people with Down syndrome will go extinct, but they may become vanishingly uncommon. One of Revive & Restore's online graphics shows extinct and threatened creatures as shadows in a landscape. I imagine a parallel image, showing silhouettes of people with genetic conditions:

Down syndrome, hereditary deafness, achondroplasia. My daughter, the gaur, the bucardo.

For many, I think, More Mammoths/Less Down Syndrome is a win-win, a desirable outcome. The pitches, in any event, are linked: the same people who argue for resurrecting lost species tend to argue for eliminating "genetic disease." Whether Down syndrome goes the way of the dodo depends, then, on whether it is considered a disease, whether it is seen as part of diverse humanity or simply as a problem we would be better off without. That no Eugenic Records Office or AES is vested in reducing the numbers of people with Down syndrome, that no coercive government agency sets current reduction targets, is not as important as it seems. NIPT is sold to individuals, and sold with a rhetoric of individual choice. But cultural values influence individual decisions, which are, in turn, multiplied by technologies into population effects.

So perhaps, one day, American fitter families will leave their future firesides, traveling on the Siberian Adventure Tourist Package to see hairy elephants tromping the permafrost. Maybe passenger pigeons will darken the skies above the subdivisions where the New England forests were, raining feces on the tract homes of genetically healthy families. Maybe 4-H will add new categories to its Small Animal competition: Best De-extinct, Best Engineered. There will be Facebook posts: *McKenzie's Passenger Pigeon won Best of Breed!!!#DIYBIO#DeExtinction#SoProud....* There will be supermuscular cavies, meat rabbits the size of Labrador retrievers, roosters with perfect dun-colored dry fly capes so long you could tie ten Pale Morning Duns with a single feather. If these come to pass, I am less concerned about the animals being shown than the children doing the showing, about who will be present and welcome.

4   On Our Screens

When I began studying the online advertising for the new, noninvasive prenatal tests, I quickly saw that the ads implied a different vision of family, disability, and American identity than the one I had come to hold. I don't oppose the tests, which many have found useful. But the ads—the *fact* of ads, as well as their substance—are deeply problematic, not only for people with disabilities, but also for the women they target.

To say that our predictive abilities are exploding is an insult to explosions. Prenatal diagnosis began, like a child in elementary school, by counting to three: in amniocentesis, chromosomes were isolated from fetal cells sampled from amniotic fluid, then arrested while dividing, stained, and photographed under a microscope. The printed photograph was snipped apart, chromosome by chromosome. The chromosomes were carefully lined up and pasted onto paper in tidy pairs. A *karyotype*: a future life, in columns and rows. The project of a good child. Our chromosomes typically come in pairs. If there were three copies of the smallest chromosome—the twenty-first—that meant Down syndrome.

As prenatal diagnosis has grown up, it has put away childish things, leaving scissors and paste behind. Arithmetic has yielded to algorithms. With NIPT, blood is drawn from the mother, and the labs separate free-floating DNA from the sample, separate the placental DNA from the maternal, and examine the placental DNA, using proprietary algorithms, to discern the likely presence or absence of a range of chromosomal conditions in the fetus. Though the web pages emphasize the tests' accuracy, they are screening tests, not diagnostic tests like amniocentesis or chorionic villus sampling (CVS)—a fact not always understood by either patients or health-care providers. (For this reason, some prefer the acronym NIPS, for noninvasive prenatal *screen*.)

Amniocentesis is diagnostic but has a slight risk of miscarriage. Other screening tests (ultrasound, for instance) have no miscarriage risk but cannot identify Down syndrome with diagnostic certainty. Hence the search for an early, safe, diagnostic prenatal test. If you believe the hype, if you read the online pitch (but not too carefully), that test arrived in 2009. Then described by Harry Stylli, the CEO of Sequenom, one of several companies then ramping up their tests, as "the Holy Grail of testing," NIPT promised the best of all worlds: safety, accuracy, and early results. "This is just a safer, more precise test," Stylli told the *Globe and Mail* in February of 2009. "It is going to save women a great deal of anxiety."

A few months after his confident pronouncement, Stylli was ousted as CEO. Sequenom's internal investigation had revealed that their new Down syndrome test, SEQureDx, was not nearly as accurate as it was claimed to be and that data had been "mishandled." The sample sizes were smaller than claimed, and the tests were not "blinded"—that is, the scientists knew which sample was which. From the Securities and Exchange Commission (SEC) report:

> January 29, 2009 Form 8-K (as amended on February 6, 2009). Stated that Sequenom had performed blinded studies, and that the Test had correctly called all but one sample, which was a false positive. In fact, the Test had been run on an unblinded basis. On a blinded basis, the Test results included multiple false positive and false negative results.

I note this recent history as a reminder that NIPT is a for-profit product, one context critical for understanding the persuasion attached to the test. The need to appeal to consumers, not to mention compete with other companies, influences the vision of the good life evident in the online ads, the contented, model-perfect moms in their third trimesters, the apple-cheeked infants; it also influences the account of the conditions tested for, which—though strenuously neutral in tone—emphasize testable conditions in terms of risk.

To write about prenatal testing, as the father of a child with Down syndrome, requires care, a respect for the difficulty and privacy of reproductive decisions, and an awareness of the things I can and can't speak to. Like many parents of children with Down syndrome, I identify myself as prochoice, prefer not to discuss abortion if I can help it, try to separate the choices made within my family from the choices made by others, and treat a woman's reproductive decisions as hers alone, an arena where my views are profoundly off the point. For this reason, I don't really have anything to say about the ethics of selective abortion. I can't imagine myself telling any woman *not* to make use of a test like NIPT, which many have found useful; nor would I presume to tell anyone what to do with the test results. What interests me is the climate of assumption: the ideas and information that feed into the choice, the ideas we have about people who are legibly different, and the way these get turned into the stories on our screens.

* * *

On January 28, 2013, the *Today* show profiled an expectant couple, Jason and Robin Vosler, who had recently opted for a brand-new prenatal test: Sequenom's MaterniT21 PLUS, a noninvasive prenatal screen.

Though nominally a human interest story, the segment was functionally an infomercial. The couple testified to the test's value, as did NBC's chief medical editor, Ms. Vosler's ob-gyn, and the chief medical officer of Sequenom. No competing test was mentioned, and MaterniT21 PLUS was compared favorably to invasive diagnostic tests. The couple shared their results on air: In a much-blogged-about exchange, Matt Lauer said, "Let's get to the good news," and asked Ms. Vosler about her test results. She replied that the test was negative, and that the family was "safe." The friendly, onstage exchange exposes common assumptions: that Down syndrome—and, more broadly, disability—is a danger to the family, and that Down syndrome should be understood mainly as a "risk" or an "abnormality," the main descriptors used in the segment. These impressions may be reinforced by the test itself, inasmuch as it tends to associate Down syndrome with more severe conditions. That association is audible in Ms. Vosler's description of the test: that it looks for "three different kinds of Down syndrome."

In fact, the test screens for three different *trisomies*: trisomy 21, or Down syndrome; trisomy 18, or Edwards syndrome; and trisomy 13, or Patau syndrome. Their effects are radically different. People with Down syndrome, to be sure, face an increased probability of a range of health difficulties, the most common being heart defects (40%) and gastrointestinal disorders. (They also have a greatly reduced risk of solid tumors, including lung cancer, breast cancer, and cervical cancer.) That said, people with Down syndrome have a life expectancy of around 60 years. Babies born with trisomy 13 or trisomy 18, with rare exceptions, live less than a year, with a median survival measured in days. People with Down syndrome attend school, and some live independently. In other words, besides the common fact of a trisomy, the conditions are not alike.

It may be unfair to expect nuance from a morning show. Even so, the program stayed remarkably true to industry talking points. Nancy Snyderman, the chief medical editor of NBC News, reiterated the central marketing message of NIPT, the one built into the label *noninvasive*: that the test is safe and accurate. When Lauer, turning to Snyderman, emphasized that amnio poses a danger of miscarriage, Snyderman agreed, shaking her head regretfully: "it does." She did not explain that women who receive a positive result will need an invasive test to confirm the diagnosis. Snyderman's deployment of medical expertise was critical: she not only burnished the test's reputation, but joined that message to a larger message about science. To do so credibly, of course, she needed to appear neutral. Offered up a softball by Lauer—he asked if she was "in favor" of the test—she replied quickly, "It's not that I'm in favor of this, it's that this is the next scientific step. It was going to happen."

Snyderman's implicit avowal of objectivity aside, her words were hardly neutral. In context, to say that something is "scientific" is a powerful and misleading argument. NIPT is significant not as science, but as technology. The existence of cell-free DNA has been known about for years; what's new is more accurate, for-profit detection. But the pitch for a specific biotech application is typically accompanied by a pitch for science and progress in general.

Closely related to that pitch is an argument for inevitability: *it was going to happen*. But to say that a new technology is inevitable, because it's progress, is a deft approach to persuasion. If your audience accepts this contention, then there *is* no debate. Of course, to say that something is inevitable leaves out the point that as members of a deliberative democracy, we can *decide* whether to adopt new advances: that the future is not something we simply accept, that happens to us, but that we can and should help to shape; that we are not merely consumers to whom technology is explained, but citizens who can think about the role of technology in society.

Robin Vosler seems to have been chosen by *Today* to deflect controversy—her intention, whatever the test result, was not to abort, and had there been a positive result, she would have sought out the specialists needed to help her child. Nancy Snyderman emphasizes this fact, and follows with a straw-man argument: "Critics will say, hey look, this is a way to find out early and then abort because we want the perfect baby." In this telling, even the ethical issues are simple: it's the appealing, telegenic prospective mom—who just wanted to be prepared, and who wasn't going to abort anyway—versus the overwrought, anti-science "critics." Snyderman's "critics" are fictional, a caricature, but even so images of "the perfect baby" are central to the marketing for the test, as is an emphasis on how early the test can be performed.

The *Today* segment concludes with a strange moment. Going into the show, the Voslers knew that the test results for Down syndrome were negative, but didn't know the baby's gender. That's revealed live, before the studio audience. The couple is handed a white box held shut with green and gold ribbon. They untie it, finding a pile of boy-themed infant swag. The dad, until then silent but smiling nervously, pumps his fist and says, "Yes!" The mom, more measured, asserts that she's "definitely" happy.

Reality TV, from *Survivor* to *The Bachelor* to *Dancing with the Stars*, depends on contrived ritual. But even by those standards, this one is bizarre: the white box as proxy womb, the future child transmuted into gendered gifts, a ritual of forecast and offering, birth and consumption, and cheerfully surrendered privacy. In a world of targeted distractions, this one is near genius: it translates a new, unfamiliar technology to a human scale and renders it in familiar terms. Nice people, opening a present.

It's a persuasive bit, and it has several key features worth noting. First, it humanizes the technology, translating it into a familiar, easy-to-imagine interaction. Second, that interaction is both value laden and emotionally charged: health, family, and a future child are all at stake. Third, it abstracts that interaction from social context: it occurs, as it were, on an empty stage. It has story's emotional resonance, without the emotional complexity—and in a world where we are all struggling to make sense of new technologies, that very simplicity has its own appeal. Like most reality-TV rituals, this one offers closure. A dancer is crowned; the poor family gets the dream mansion; a future wife is selected with a rose.

And yet the complexities keep spilling over the tidy boundaries of the form. In the online version of the story, the test results are described as "good news on two fronts." It's troubling to see the "good" news of a fetus not having Down syndrome linked to the "good" news of that fetus being male, because whether these are "good" are not only matters of biology, but matters of value. For this reason, the presentation of a single story is misleading. Who could begrudge the Voslers their happiness at a boy, their wish for a healthy child? But the Voslers aren't the point. What matters is the way that pitches for new technology both absorb and reinforce societal norms. Chromosomally, the child is solidly male (XY) and typical (23 pairs of chromosomes, with no trisomies), and the family is nuclear, with two monogamous, heterosexual parents.

In the *Today* show's rough arithmetic, a syndrome is a matter of health, and gender is a matter of humanity. One is bad news, the other good. As it happens, the test can also detect conditions with atypical numbers of sex chromosomes: Turner syndrome (a single X chromosome), Klinefelter syndrome (XXY), Triple X syndrome, and XYY syndrome. In other words, the test marks deviations from typical gender while being associated, for profit, with an example of a committed heterosexual couple, welcoming a normal baby boy.

Rituals like these have no history for the people who practice it. They're invented off camera and derived from existing rituals. (This one combines Baby Shower, Gender Reveal Party, and Delivery of Diagnosis.) But all are subsumed by a ritual specific to the medium: the reality-show reveal, on shows like *The Biggest Loser* or *Extreme Makeover*, in which the human body is altered, quantitated, and unveiled. The difference is that in the genomic age, an age of biology and information, it's the genes under inspection—or the data the genes have become, processed by algorithms to expose a probable future.

The Reveal reveals a surrealism particular to our time and place. The body is measured: pounds lost, chromosomes counted. Strict gender roles are enforced—beautiful woman, handsome man, perfect baby boy—along with norms of ability and health. The whole is broadcast for profit, and multiple corporations benefit, through advertising and product placement. Responsibility is placed on the individual: to lose weight, get better hair and clothes, purchase a prenatal test. Happiness is equated with certain kinds of bodies and not others. Society's role is not to accommodate different sorts of bodies; the government is nowhere in sight; the corporations are benevolent. The experience is only meaningful because it is watched. We are consumers and voyeurs: our job is to watch, judge, shop, and share. The gift, the secret, is information.

To characterize a child with disabilities as a gift is to be dismissed as deluded, to be considered unrealistic by those unfamiliar with the reality. In fact, to say that a child is a gift doesn't go far enough. Doing so still renders the child as an object of meaning, not a source, and neglects the relationships in which both children and parents discover meaning. Gail Landsman, in her study of mothers of children with disabilities, writes that many mothers arrive at a different interpretation:

> Counter to their expectations, these mothers did not get what they worked for; the "product" is judged below standard. Yet, if a woman later redefines her child neither as a product in the commodity market nor as a gift from God but rather as a giver of gifts, she raises the value of her child beyond that of the "perfect" child she had once anticipated and strived to obtain. Representation of the disabled child as giver is particularly subversive in American culture wherein the personhood of those with disabilities is diminished in large part because such individuals have been viewed not only as incapable of giving, but as relying upon the gifts of others.

The resistance to misconception, no less than the misconception itself, takes metaphorical form. Responding to a dominant narrative, parents craft their own. Seeing a child as a giver of gifts is also significant because it reframes the child as having something to contribute, yet leaves the contribution open. It asserts a meaningful relationship without grounding it in achievement, intellect, or performance.

* * *

As a public ritual of normalcy, defect, and family, the *Today* show segment is a distant descendant of the telethons—the marathon TV fund-raising appeals, each centered on a disease, that began after World War II and ran through the 1990s. Jerry Lewis's telethon for muscular dystrophy was the best known, but others centered on cerebral palsy and childhood cancer. There's a crucial difference, though: the *Today* show offered a drama of a tragedy averted, while the telethons offered the spectacle of disabled children as living tragedies to prevent.

In his classic history *Telethons*, Paul Longmore showed that the charity drives were also media rituals, in which the disabled child was the linchpin of a system of meaning. With its gamely optimistic but tragically afflicted children, telegenic middle-class families, prominently featured corporate benefactors, and heaping helpings of schmaltz, the telethons raised millions for research and assistance. Longmore acknowledges that the telethons had benefits, but that these came at a cost: the price, for people with disabilities, was a "restrictive and depersonalizing social identity." While the telethons made people with disabilities more visible, the visibility was on restricted terms: they were defined "as objects of benevolence rather than as autonomous subjects":

> Normal people gave alms; invalids took them. They no longer squatted on street corners proffering tin cups, but they were beggars just the same. It was their role to receive these gifts gratefully. They existed to supply nondisabled people with an occasion to exercise charity and thereby reaffirm their own normality.

The variety-show format, Longmore argues, was animated by narrative tropes drawn from "sentimental literature and Victorian fiction," including "unmerited affliction through illness or accident, the suffering of innocent children, and emotional excess." The poster child was a real-life descendant of Dickens's Tiny Tim, an appeal central to the telethons' fund-raising goals. Longmore quotes a publicist for the March of Dimes: "[T]here is an emotional appeal with a child with crutches and braces—that's going to bring the money in." But as these children came of age, many began to question the uses to which their stories had been put and to tell a story on their own terms.

In her memoir *Poster Child*, Emily Rapp tells the story of growing up with a prosthetic leg, and—at six—reveling in the attention she received from appearances for the March of Dimes. As an adult, though, she understands the message beneath the effusive welcome: "My six-year-old grin beamed beneath the March of Dimes motto: 'Help Prevent Birth Defects.'" A superficial message of acceptance conceals, and ultimately supports, a deeper message of elimination. "At church functions, at rodeos, and at other community venues and events," Rapp writes, "I spoke to crowds both large and small about how normal my life was and how happy I was—all in an effort to raise awareness and money for an organization designed to fund research that would prevent congenital birth defects similar to my own."

Rapp's take on disability is complex and unsparing. She remembers, for example, seeing intellectually disabled students in special ed and being grateful, as a child, that she wasn't "one of them." She is frank about the difficulties, both social and physical, of having a prosthetic leg, even as she adapts—she becomes, for example, an expert skier. But most of all, she shows the way that her disability is inseparable from every other aspect of her life. Her struggles with appearance, achievement, food, spirituality, sexuality, and family are all interwoven with her understanding of disability, a fact of the body whose meaning is fluid. It depends on technology (improvements in prosthetics improve her quality of life); on people around her, whose responses range from rejection to provisional acceptance to unquestioning understanding; and on her evolving perspective, in which she awakens to fierce societal norms of embodiment. She realizes the extent to which she has internalized those norms and the damage she has done to herself in the pursuit of normality.

Rapp's memoir not only shows us that disability is more complex than supposed; it also questions who gets to say what disability means, and how. To be a poster child is (by definition) to be on display, with a meaning assigned to one's life; to write a book revises the terms of disability, claiming a public self on her own terms. She moves, in Longmore's terms, from object of benevolence to autonomous subject; against the telethon's script of separation, she scripts a conditional belonging, or a process by which belonging might occur.

*   *   *

In early 2013, I'd found myself spending a lot of time at Today.com. I'd streamed the Sequenom segment at Today.com/Health, and just a week before that, I'd written about a feature at Today.com/Moms, which offered an utterly different vision of disability and family, in which Down syndrome was not a threat to keep a family safe from, but the core of a family's happiness. The story was titled "Waiter Hailed as Hero after Standing Up for Boy with Down Syndrome." It was a write-up of a news report from KPRC-TV, an NBC affiliate in Houston. According to the article, the Castillo family—Eric, Kim, and their 5-year-old son, Milo, who has Down syndrome—were having dinner at Laurenzo's Prime Rib when a family nearby asked to be seated elsewhere. The waiter, Michael Garcia, tells the story:

> I heard the man say, "Special needs children need to be special somewhere else." My personal feelings took over, and I told him, "I'm not going to be able to serve you, sir."
>
> "How could you say that?" Mr. Garcia said he asked the man before he left the restaurant. "How could you say that about a beautiful 5-year-old angel?"

An angelic underdog, a bully with a catchphrase—*be special somewhere else*—a lone hero, and a Texan showdown: How could the story *not* go viral? In a *Field of Dreams*-like ending, business was reported to be booming at the restaurant.

Like the Reveal, the Feel-Good Story is an inflexible form, sentimental and rigid. It is an algorithm to produce emotion, a fable to produce a lesson. On the surface, this story is no different, particularly in its emphasis on Milo's cuteness. (Beneath one smiling picture is the caption "Milo Castillo has lots of friends in preschool and loves to give hugs"; beneath another, "His mom takes him out to restaurants frequently and says he's very well behaved.") So deployed, the facts enhance the drama: innocent victims are more sympathetic. Since Down syndrome is commonly associated with sweetness, innocence, and placidity anyway, the stereotype fits easily into the form.

And yet, just as with the scripted Reveal, the situation's realities spill over the form's boundaries. If you look beyond its tidy ending, you can learn a lot about the conflicted place of Down syndrome in the world. The casual insensitivity of the patron's remark, while not always expressed this blatantly, is all too common. At the same time, the waiter stepped up in a rare way. The casual abuse of the term "hero" does not mean there's no such thing as heroism; as Ms. Castillo pointed out, Mr. Garcia put his job on the line. He knew the Castillo family, who are regulars at the restaurant, and he acted in the name of connection, denying the very separation—between "special" and "normal"—that the unidentified patron implied. Ms. Castillo's characterization of her son also broke through the prescribed roles of the story. Her description is loving and positive, yet she emphasizes ordinary qualities as well as angelic ones. She said her son likes to give hugs, but is also occasionally "obnoxious which, like any other 5-year-old, he can be."

She also noted that Milo has friends "both with and without Down syndrome." That detail stayed with me, not only because it is true to my experience but also because it suggests how far we've come. During decades of routine institutionalization and sterilization, a shared preschool—let alone friendship or being a valued regular customer at a Houston restaurant—was not on the table. If there were a slogan for those grim decades, "Let them be special somewhere else" would sum things up pretty well.

It's moving to me that Milo Castillo's story takes place in a restaurant, given the civil rights history of lunch counters and water fountains. What's a more visceral expression of belonging than where we're allowed to eat? Where else would manners and rights intersect? As presented in the television report, the opinion voiced by the unidentified patron—"be special somewhere else"—is seen as a shocking instance of rudeness. It is. But more significantly, it's the sentiment that, when widespread, makes it possible for people to lose rights in the first place.

Dehumanizing practices depend on dehumanizing ideas. Instead of seeing the scene in the restaurant as a story with a heroic resolution, we might see it as a tiny fraction of an open-ended conversation about the meaning of disability. Beneath the human interest story, in other words, is a question about who counts as human, and the conversation—carried out in words and actions and images, in person, in print, on TV, and online—is a key part of our contradictory, perpetual *Today*, where one day a child with Down syndrome is a loved family member to protect, and on another day, someone to protect a family from. In one, happiness implicitly hinges on the absence of a child with Down syndrome; not being pregnant with one is "good news," and because of a test, a couple is "safe." In the other, happiness hinges on his presence and on the presence or absence of welcome. You couldn't invent a starker illustration of our attitudes toward Down syndrome, and disability in general—or the difficult questions coming our way, as our increasing acceptance of people with disabilities collides with increasingly accurate prenatal tests.

Those tests are the subject of conversation, but they also drive it. The tests do not merely detect genetic variation. They *mark* what is abnormal, shaping our understanding of it by reshaping the category of abnormality. At the same time, they allow our ideas to be translated more easily into populations, and the communication associated with the tests (pamphlets, online advertising, *Today* show features) drives specific ideas about disability and family. Though we use the word *normal* as if we agreed about what it meant, the word's implications change rapidly, year to year. In this way the technology becomes woven into our lives. It defines what is conventional and unconventional, in human bodies; as a practice, it not only lends those conventions power, but becomes a convention itself, something we do as a matter of course, a norm that defines a norm.

There have always been people with anomalous bodies, behaviors, faces, abilities. What their lives are like depends in part on their bodies, in part on how their embodiments are understood. As Gail Landsman notes, there is no universal response. Some cultures have practiced infanticide, leaving disabled infants outside to die of exposure. In others, they are welcomed. Landsman cites research on the Cuna Indians of Panama, where men who inherit albinism become night fishermen, and on the Punan Bah of Central

Borneo, where "physically and mentally impaired children live as members of extended families, share in age- and gender-appropriate household work, and are acknowledged both as humans and as persons."

It is possible to imagine a society in which people with intellectual disabilities fully belong. There may have been such societies in the past, and there may be others in the future, and in the meantime we and the other families I know tend our improvised spheres: fragments of belonging, like memories or predictions. There's no better time and place to raise a child with Down syndrome, and our lives are luckier than many. But our path still goes against the grain.

* * *

Our understanding of Down syndrome is endlessly complex, but it is always wedded to the technology of the day. The ability to stain and count human chromosomes, combined with amniocentesis and legal abortion, meant that Down syndrome could be detected and prevented. As a result, the meanings associated with Down syndrome changed radically. Down syndrome came to stand for other conditions detectable in utero, however dissimilar. If Down syndrome is emblematic, that is, in part, an artifact of technology. We stigmatize what we can see, and people with the condition are instantly recognizable at two levels of magnification: their features are distinct, and chromosomes are easy to see under the microscope. The people have absorbed stigma in the way chromosomes absorb stain, but the expression of stigma reflects time and place.

In May of 2012, the comedian Margaret Cho appeared on the TV show "Watch What Happens Live." Discussing her wish to become a mom at 43, she said, "My period comes like twice a month. My eggs are jumping ship... seriously, they're like, 'the last one out's a retard.'" She continued, "I get worried about that, as an older woman, I don't necessarily want to have a retard." The audience laughed; so did the host, Andy Cohen, who also hid his laugh behind his cue card, saying, "You can't say that." Ms. Cho appeared surprised that anyone would object and also said, "You want your kid to have the best chance at life... I'm trying [to get pregnant]. It's hard for a lesbian." Though Cho does not specifically mention Down syndrome, it is clearly the source of her concern. It is unspoken and emblematic, invisible and central. The outcry over the joke resulted in two public apologies on her website: the first a brief, pro forma celebrity apology and the second an extended, self-flagellating confessional and plea for understanding.

While it's true that open bigotry is less common than it used to be, the absence of hostility is not the same as the presence of welcome. The word "retard" is common in the schoolyard and on our screens: while our taboos about difference shift with incredible speed, slurs about intellectual disability often get a pass. From the movies *Tropic Thunder* to *The Hangover*, from Rahm Emanuel to Ann Coulter, "retard" poses as edgy, not bigoted.

Jokes like this, as Theresa says, are only funny until you know someone with an intellectual disability. Few people do: the laughter at Cho's joke shows how segregated our society is. Despite the progress of recent years, despite the occasional Special Olympics Unified Team, people with Down syndrome, and intellectual disabilities generally, are effectively separate from the mainstream of American life. An audience full of people who cared for people with disabilities, who counted them as friends and family members, would not have laughed. That people *with* intellectual disabilities might be in the audience seems not to have occurred to Cho at all.

Cho's joke not only demonstrates longstanding anxieties about children, health, and disability: it shows how quickly we embrace and normalize new technology. It's only recently in human history that genetic conditions could be diagnosed in utero or that the correlation between maternal age and Down syndrome became known. That these facts could sink to the subconscious and resurface in a joke shows how quickly our cultural understanding of family, reproduction, and disability can change. Cho's understanding of disability and motherhood was inseparable from the technology available.

Though Cho's apology—or, at least, the fact that she did apologize—is laudable, it is also shot through with the same misconceptions and anxieties about disability that fuel her original routine. The central assumption—that disability is a burden for a family—remains:

> Know that the children of the world, especially those differently-abled kids and their brave, ever noble parents and families, who have it hard enough to begin with, deserve much better than me and my idiotic need for approval in the form of nervous laughter.

Left unquestioned is the belief that it is a bad thing to be disabled and for a parent to have a disabled child. Those parents are "brave" and "ever noble," but not real. "Idiot" is the "retard" of another century, a word rooted in both insult and diagnosis: the original label for Down syndrome was "Mongolian idiocy." The irony of Cho's saying it was "idiotic" to use the word "retard" illustrates the durability of our obsession with intellect, and the way our assumptions about intellectual disability pervade and structure our thinking. Those assumptions, like the people themselves, are present and unnoticed, in plain sight: in online comment sections on every topic, the words *idiot, retard, moron,* and *imbecile* all echo uncounted times each day, traces of early twentieth-century American eugenics, words that were neologisms once. The history is present and active in our daily lives, whether we notice it or not.

For Cho, the sense of disability-as-burden is deeply embodied. Her fear, she writes, is that her years of mistreating her own body, "so many years of acrimony and anorexia, hangovers and overtime, vicious colds and heart-aches I have never allowed myself to recover from," would somehow result in a disabled child: "I fear my body will have the last word, and instead of penalizing me only, it would hit me where I really live, in the body of my preciously abstract yet to be conceived child."

Down syndrome is correlated with maternal age. It is not correlated with stress, anorexia, alcohol abuse, bad colds, or heartaches. Taken together, Cho's routine and apology offer a mash-up of new and old ideas: the new understanding of genetics and maternal age; the slur "retard," itself a per-version of the dated noun *retardation*; the ancient idea that disability is a punishment for sin, though updated for the New Age, where a disabled child is punishment for insufficient self-care.

Cho also repeats her idea, expressed on "Watch What Happens Live," that she wants to give a child "the best chance":

> I think life is hard, and this planet is an especially unforgiving one—spinning thoughtlessly and carelessly on its axis without regard for humanity and all those who suffer daily from the dizzy—and if I want to drag someone else into this mess of a world—an innocent soul, a mere baby bystander—I want to give that kid the best chance possible.

Cho's apology is far from perfect, but I take it as sincere. Her routine was abhorrent, her apology troubled, but neither changes the fact that personal decisions about reproduction are often extraordinarily difficult, and worry over future children—who are, as Cho says, "preciously abstract"—is more than understandable. That worry, as we'll see, is one target of the market-ing for NIPT, which presses on anxieties about risk and health, even as it presents images of ideal families. But Cho's apology expresses another assumption unconsciously congenial to corporate interests: though she rightly perceives that the world is often cruel, and that one's chances are helped by being smart, her solution to this problem is purely individualistic. It is not the world that needs to be welcoming to people with disabilities; it is the *mother's* responsibility to give birth to a child with "the best chance."

* * *

The kerfuffle over Cho's comedy centered on a single joke. That's what I assumed she was apologizing for, and that's the clip I watched at the time. But as I tracked down references for this book, I found a second clip from the after show, when guests take audience questions. Riffing on the word "retard" again, Cho makes the connection to Down syndrome explicit:

> **Andy Cohen [reading from card]:** Stephanie D. wants to know if you've ever regretted a joke that you made on stage.
>
> **Margaret Cho:** Umm…I am really sorry that I said "retard" about twenty, thirty minutes ago. [laughter] I've been thinking about it this whole time and I realize I'm going to have to make a Special Olympics presentation or something, and I know that is really—
>
> **Andy Cohen:** Send a tweet.
>
> **Margaret Cho:** I feel bad, like I'm going to have to go all Tracy Morgan and do, like, work with whatever the GLAAD of retards is. So I have to find out and get in touch with these people…I mean, like, I don't know, I feel bad about it, because I know I clearly look like I have Down syndrome, so I have issues with retarded people, cause people clearly mistake me for one. Especially when I drink. Why do you have alcohol on the show?

As I've written elsewhere, this confusion of ethnicity and disability is familiar: it's been a part of my story ever since Laura came along. Because my mother is Japanese, the doctors hoped that the shape of Laura's eyes might indicate her heritage, not her chromosome count. People with Down syndrome do not actually look Asian at all, but that nineteenth-century confusion lies beneath the label *Mongolian idiocy*, the early, uncertain days when we awaited the results of genetic testing, and Cho's joke. *I know I clearly look like I have Down syndrome.*

That discarded label sprang from a belief in the hierarchy of race: John Langdon Down, the Victorian physician and medical superintendent of the Royal Earlswood Asylum for Idiots, thought that a few distinctive-looking residents in his care had "degenerated" in the womb, dropping down a hierarchy of races from Caucasian to Mongolian. Down's theory—which claimed the existence of other ethnic categories of idiocy, including

"Negroid" and "Malay"—imposed boundaries on disability's fluidity and nuance, answering individual lives with fixed types.

Though Margaret Cho's comedy—*I know I clearly look like I have Down syndrome*—comes from an utterly different worldview, her radical devaluation of intellectual disability is similar. But where Down fused ethnicity and disability, Cho separates them. In doing so, she constructs an idea of diversity in which disability is excluded, ridiculing the idea that disability might count as a political category of identity. *Whatever the GLAAD of retards is.*

* * *

To be intellectually disabled is to have your life synonymous with an opinion not worth listening to: on Facebook, in every comment section, in conversation, that's what the words *idiot, moron,* and *retard* imply. People with intellectual disabilities, besides being among the most despised minorities in our culture, are cast in a harsh light by a society that prizes intellectual ability and accomplishment. Negotiating our text-heavy, Information Age democracy requires an unprecedented degree of literacy and technological ability. In work, in education, those abilities are heavily incentivized. Indeed, our educational system encourages us to equate intellectual performance with self-worth, to motivate ourselves by seeing ourselves as our grades and accomplishments.

For this reason, the divide between able and disabled can be seen in terms of interpretive power. That means that description is an ethical act, and that people able to navigate a complex conversation, both on- and offline—especially those with an audience, like comedians, TV hosts, or writers—have responsibilities to the people they describe, which go well beyond avoiding outright slurs. For those who are skilled with language, at ease with abstraction, and able to process, retain, and manipulate large quantities of information, a key question is how to imagine the people who either lack those abilities or have them to a lesser degree: those who, in a competitive economy, find it more difficult to perform their value. For those who are unusually visible, the question is how to imagine the invisible, the residents of the margins.

So it may make the most sense to see Cho's riffing on "retard" as an economic miscalculation. Given a rapidly shrinking pool of comedic targets, jokes about intellectual disability offer a way to seem edgy without derailing a career. From that perspective, Cho's joke was a bid for continuing status in an economy of attention, and her apologies, however sincere, were also a bid to defend and maintain that status. But the backlash suggests that a change is already here and that the joke was a misreading of the zeitgeist, and one swiftly punished in the age of information.

If the zeitgeist is changing, it is increasingly because people with intellectual disabilities are beginning to speak up. Replying to Ann Coulter, who in 2012 tweeted a description of President Obama as a "retard," John Franklin Stephens published this in the *Huffington Post*:

> I'm a 30 year old man with Down syndrome who has struggled with the public's perception that an intellectual disability means that I am dumb and shallow. I am not either of those things, but I do process information more slowly than the rest of you. In fact it has taken me all day to figure out how to respond to your use of the R-word last night …
>
> Well, Ms. Coulter, you, and society, need to learn that being compared to people like me should be considered a badge of honor.
>
> No one overcomes more than we do and still loves life so much.
>
> Come join us someday at Special Olympics. See if you can walk away with your heart unchanged.
>
> A friend you haven't made yet,
>
> John Franklin Stephens
>
> Global Messenger

More recently, Mr. Stephens responded to comedian Gary Owen, who in a Showtime special made fun of his disabled cousin, disclosed that she had an STD, said "I didn't know retarded people had sex," and ridiculed the way Special Olympics athletes run. Stephens wrote,

> What bothers me is that Gary Owen and at least some of his audience found it easy to believe that no two things could be less likely to belong together than a person with an intellectual disability and sex. That is the real tragedy here. Society, even people who care most for us, simply can't see us as completely human enough to imagine that we have the same desires and needs as the rest of you.

Owen eventually met with a group of disability advocates (including Stephens), apologized, pulled the offending segment from his Showtime special, and promised not to use the word "retarded" in future routines.

This kind of humor enacts the separation on which it depends: it divides the world into those who do and do not get the joke, into those who laugh and those who are laughed at. Wit, at the expense of those considered "witless." That separation is the real problem. It's easy to agree that these routines are offensive; we should also be asking why it is that so few people with intellectual disabilities are part of our public life in the first place, why it is still noteworthy to see a man with Down syndrome at a comedy show, emerging from a voting booth, a brewpub, or a bookstore, and why it is still vanishingly rare to see people with intellectual disabilities not merely represented in the debate, but representing themselves.

That's one reason Mr. Stephens's reply was so effective. It's not just that he was thoughtful in a way people with Down syndrome are not commonly supposed to be capable of, or talked about sex and pleasure in the context of intellectually disabled adults' lives, or was as principled as the comedian in question was crass. It's also that he spoke up at all. In so doing, he not only won the debate, but questioned its structure, wherein able-bodied people speak to other able-bodied people *about* disabled people, but disabled people are rarely heard.

5   The Fine Print

When I first began researching the websites pitching NIPT directly to women, I happened on an explanatory graphic: a simple cartoon, readable from left to right. The first image was a vial of blood; the next, fragments of DNA (green for the mother, blue for the fetus); after that, an image of sequencing data and a picture of a computer; and last, a sheet of paper, evidently reporting test results. From blood to paper, with an algorithm in between: the image crystallizes our time, with its incessant, computer-mediated translations, genes to bytes, blood to text.

Every website had its explanations. Shadowless pictures of arms, needles, monitors with science stuff on them. There was usually a person, or part of one—an arm, a woman in a chair, a vial of blood—and usually a computer and little double-helix fragments that seemed native to both, that seemed equally at home in vivo and in silico, as if the molecule could pass easily from one realm to the other.

Because the tests predict the nature of future children, it seemed appropriate, somehow, that the images seemed designed *for* children, like the Social Stories from Laura's preschool, the four-panel, laminated clip-art cartoons used to explain the day's routine: *I hang up my coat, I sit on my carpet square for circle time*, and so on. But then the ads *are* social stories, guiding the consumer from information to decision. At this writing, for example, Sequenom's home page is dominated by five clickable circles, each a step in the process. From left to right: *Explore our genetic tests, talk to your doctor, find blood draw location, understand billing and insurance, schedule results counseling.* Each circle pulses when you mouse over it, like an invitation to learn more. If you take the word *algorithm* literally—a sequence of steps, leading to the solution of a problem—the graphics illustrate an algorithm, in which the consumer is a key variable.

It's strange to write about web-based advertising in a medium as slow and clunky as a book. The websites for NIPT are updated frequently, altered overnight in response to claims and criticisms, revamped to reflect a new study, an emphasis on new disorders, a corporate takeover, a brand-new look. Soon after I began writing about Sequenom's MaterniT21 PLUS, ads for the product began popping up in my Gmail account. Shortly after a presentation I gave in New Orleans, to the Annual Education Conference of the National Society of Genetic Counselors (NSGC), some of the text I'd commented on was deleted from the website. The ads change with extraordinary speed. And yet their formal features remain the same: the presence of ideal bodies, the absence of disabled ones, and the button to request more information.

Like so much of the code we live by and depend on, from Google searches to the computers in our cars, the algorithms on which NIPT is based are proprietary. NIPT illustrates an irony in our world: we are inundated with information, so the world feels transparent, but the code we depend on is closed to us. At the same time, because the number of conditions reported by NIPT is rapidly increasing, NIPT contributes to the information overload typical of our age—and increases our reliance on experts to decipher the information's importance. And yet the websites bypass the experts most needed to make sense of the test. In a way impossible before the Internet, the sites target women directly with a carefully constructed, profit-driven presentation of the test, the conditions tested for, and women themselves. The sites do more than try to sell a product; they offer a vision of the good life, made possible by technology.

* * *

They look like models from a Cialis ad: healthy, prosperous, white, late thirties or early forties. If you were guessing, you'd say that the man is an executive in a nonmedical field, that the wife has a professional degree but scaled her career back for family, and that they drove to the office together in a silver Lexus SUV. His hobby is golf; hers is scrapbooking. You see them over the doctor's shoulder—a blurred white coat in the foreground—and they look concerned but reassured, as if they have just received good news about a solvable problem. The husband's arm is positioned supportively behind the woman's chair. There are no markers of political and religious affiliation: their story is a matter of suggestion and erasure, underpinned by the certain fact of an extra chromosome. In this way, at least, it resembles the story of the condition they are clearly there to prevent.

The image appears on Sequenom's website in 2013. Much later, I discovered that the same image, of the same couple, had been used twice at *Parents* magazine and once at *Deseret News*; it was also used to advertise the services of an audiologist, a number of fertility clinics, a company selling replacement windows and doors, and a surrogacy clinic in Ukraine. Similarly, a picture of a child in a meadow, used by Natera, was also used by the website SafBaby (in an article about Forest Kindergartens), by a Canadian debt restructuring company ("Is there a way to reduce my child support/alimony payments or defer them temporarily?"), and by a company manufacturing all-natural insect repellent bracelets. That these stock images circulate so easily between genetic technologies and alternative medicine, between medical and nonmedical selling, suggests the power of these images, and the way in which online persuasion depends on idealized images of family.

Taken together, the images used to sell NIPT articulate a suburban utopia. Call it Ariosaville, Sequenomia, Nateratown: it is American but regionally vague, an exurb of some major city or other, upscale but not out of reach—the people a little more beautiful, the kitchens a little bigger, the picture windows a little wider. Generic Americans, attractive but not recognizable, nominally diverse, like models in a Pottery Barn catalog. Fictions alive in a slipstream of fictions. Characters in an implied narrative, whose implications the ads carefully unfold.

The ads' aesthetic is Sunlit Clinical, with a strongly feminine slant. Against tastefully floral background colors, a mall-atrium sunlight falls across yoga-practicing, Whole Foods–shopping, diverse-but-white-leaning Madonnas in their third trimesters, hands lightly cupping their bellies. Men, when present, are handsome, dependable models who embrace their pregnant life partners supportively yet protectively. In the vast emotional spectrum of love, pregnancy, and motherhood, the websites occupy the narrow lavender-shaded bandwidth between Joyful and Serene. It is a demographically narrow world as well, a likely effect of the test's cost: there are pearls but few piercings and no tattoos. No one is unhappy, overweight, or nauseous. Besides the women doctors with stylishly short hair and funky glasses—a blend of Your Cool Mom and The Doctor You Can Trust—and the occasional chromosomally typical child cavorting in a meadow, there are only women that someone behind the Nordstrom's cosmetics counter would be actively happy to see.

When I traveled to New Orleans to speak to genetic counselors and industry representatives about the ads, I projected images of women from the websites, to which I had added captions:

Some Days I Feel Joyful, but Other Days I Just Feel Serene

I Have Time to Lie Around in Meadows

I Have Money and I Like My Body

I've Never Experienced Nausea or Excessive Weight Gain

Namaste, NIPS!

A Male Model Loves and Protects Me

I decided not to project the caption "I Have Ethnic Friends." Based on the feedback I got afterward, my sense was that the genetic counselors enjoyed the presentation, and that the industry representatives did not.

The women in the ads are shown in a limited number of roles: being happily pregnant; being loved in an apparently monogamous, heterosexual relationship; listening to doctors attentively; reading pamphlets with interest; and mothering a beautiful, genetically normal child. Combined with the insistent messages about empowerment and choice, the total effect is profoundly mixed: women are independent (but protected by a man), thoughtful (but always receiving information), and empowered (but dependent on technology).

But to ask about the portrait of women in the ads for NIPT is to look beyond how they are represented and to consider how they are addressed.

Ariosa Diagnostics' website was one of the first I visited when I began looking closely at the companies selling the new early-pregnancy fetal gene tests. In 2012, it mainly featured attractive women relaxing in their third trimesters. (It is difficult to find images of men on the Ariosa site.) The pitch—explicitly addressed to women, in an uncomfortable hybrid of Scientific Authority and Just Us Girls—is worth a closer look:

> As an expectant mother, you have probably been overwhelmed with information about what you should do during your pregnancy. And more than likely, you've considered prenatal testing.

This is a brief and deftly written paragraph. It offers sympathy, while separating Ariosa's information from the "overwhelming" kind, then pivots to the choice. (Notably, it frames that choice as something the mom is *already thinking*—which may be true, if machine learning has determined, from the mom's searches, that she's pregnant and may be a potential customer.) It's those *other* companies overwhelming you: Ariosa Diagnostics wants to help. And by the way, since we're helping, here's some information:

> According to the American Congress of Obstetricians and Gynecologists (ACOG), all pregnant women should be offered prenatal testing for chromosomal abnormalities.

The ubiquity of ACOG's recommendations is a testament to their persuasive power. If something is medically recommended, it is easier to see it as a matter of health, and not as a value-laden choice. What follows is instructive in two senses: it tells the consumer about prenatal testing and instructs her in how to act.

> Prenatal testing is part of almost every pregnancy. Your doctor can provide you information about trisomy prenatal testing, including why it's done, what tests are available, what they are for, and when you should get them.
> The choice is yours. So learn all you can about the tests available to you, and if you have questions, discuss them with your healthcare provider.

Brief as it is, this passage is loaded with contradiction. It emphasizes the prospective mother's choice while telling her what to do. It asserts what is technically correct—that "[p]renatal testing is part of almost every pregnancy"—while omitting the fact that *this* kind of testing is relatively new. Finally, it blurs the meaning of the word "information." In the paragraph above, "what tests are available" and "what they are for" certainly count as "information": those are unambiguously factual. However, the question of *why* that test is done is charged with value. In the same way, the web page itself blurs information and persuasion. Its tone is neutral, but its clear purpose is to nudge the consumer towards a purchase. That's why the passage, brief as it is, emphasizes physician contact: behind the neutral-sounding delivery of information is a fragmentary narrative, in which the consumer is a character, someone who has, "[m]ore than likely ... considered prenatal testing." The paragraphs attempt to pique and channel that curiosity, to move the consumer from an internet search at home to a consult in a doctor's office, to get her to the next plot point in the narrative.

In context, the phrase "[t]he choice is yours" is ironic. Ariosa's website is typical: the ads, without exception, invoke the ideal of choice, even as they undermine it. The deployment of images and words—apparently informative, but in fact persuasive—is part of a campaign to increase uptake of the test, which, as with pharmaceutical marketing, includes a pitch to medical professionals. (Websites focused on NIPT are nearly always partitioned into a "for patients" and a "for doctors" section, and genetic counselors are subject to a steady stream of marketing, including in-person visits.) Despite the rhetoric of empowerment, the ads are filled with imperatives: "print out this questionnaire," "ask your doctor," "demand the accuracy of this test," and so on.

Choice can be eroded in many ways, and the ads' portrait of women offers two: it trivializes its audience, offering them ideals of pregnancy and home as if these were relevant to a serious medical decision, as if the women *needed* these images in some way, and it trivializes the choice to test itself, presenting a citizen's decision as a consumer option.

Which brings us back to the images of home. The spacious patios, the expanses of glass, have economic overtones. They are images less of belonging than of American success. The ads extol the system they inhabit, offering a well-off world where everyone is healthy—the occasional grandmother is reliably hale—and people with disability are absent. That absence is structural, not incidental: the desire to avoid disability drives the invention of the technology, provides the technology's primary market, and haunts the pastel worlds the ads depict, where everyone is beautiful, no one is nauseous, no one worries about whether they can afford health care, and the children do not ever cry. It is an entire representational world lit by fear and profit; it flickers on the cave wall, like the shadow of disability.

It's tempting to say that the ads tell a story about women, families, disability, and technology, but this would only be partially true. Stories belong to language, and language is only part of a website's appeal: the images are more powerful. Actual stories are rare, perhaps because they are too specific. What a consumer faces instead is a bespoke assembly of fragments: videos, links, images, slogans, statistics, explanations, testimonials. There's an implied narrative, or the *materials* of a narrative, whose setting is a nameless upscale suburb, whose characters are the radiant, third-trimester moms featured in the ads. But it is the consumer's job to complete the narrative: literally, by navigating the information and appeals of the site, and figuratively, by inserting herself into it, by becoming a satisfied user, like the ones depicted in the ads. As a consumer of digital media, her job is to piece raw materials together into a story, to organize disparate fragments into a possible future. To do, in other words, what the test does.

Ads for prenatal tests, like ads for lawn mowers, soft drinks, chain restaurants, and all-natural insect repellent bracelets, feature images of the good life: happy families, living well. Unlike those products, prenatal tests have the potential to influence what the families in question will look like. The path of influence is paved by the digital: to the extent that the ads help connect the consumer to the company (through a "get more information" button), they remove barriers between the consumer and the use of the technology—which is itself digital, a tool to break down barriers between blood and paper, between genes, bytes, and words.

*   *   *

Health and persuasion have long been interwoven. In *The Attention Merchants*, a history of modern advertising, Tim Wu notes that patent medicines were key to the rise of commercial persuasion. Those "medicines" generally ranged from worthless to poisonous, but the promises made on their behalf were as extravagant as the promises attached to biotechnology today, the lists of diseases and disabilities as wide-ranging. Wu writes, "What more basic and seductive human wish than to be cured of one's infirmities?" Some even "[promised] immortality, deliverance from the greatest fear of all."

In time, crude promises of health yielded to more sophisticated approaches. The ads for Listerine—a floor cleaner, repurposed as mouthwash—offered a new word for bad breath: *halitosis*, a clever coinage that drew on medical authority while tapping into fears of embarrassment. (The fictional star of its ad campaign was "poor Edna," a beautiful woman in her thirties—but an old maid, because of her terrible halitosis.) Another campaign, by the pioneering ad writer Helen Lansdowne, pitched Woodbury's Soap with an image of a beautiful woman, held by a "dashing man": "unlike the traditional ads," writes Wu, "which offer a cure for a problem—new snake oil in old bottles—Lansdowne's advertisement holds out the promise of a better life. It sells the reader on herself, a new self, better than the old."

With this approach, the advertisers both sold products and reinforced ideas about which people mattered. They dangled ideals for people to almost reach, in which the values of the industrial society itself were reinforced. "Poor Edna," the subject of the Listerine ads, was barred from marriage and all it entailed: bad breath kept her from the good life. Conversely, the successful user of Woodbury's Soap had found happiness. There is an almost eugenic undertone to the ads, an understanding about who was fit to reproduce.

These pitches worked by targeting "subconscious anxieties" to get people to want things they hadn't thought to want:

> As consumerism grew, it also became possible to sell products solving problems that were hardly recognized as such, let alone matters of life and death. Demand was engineered by showing not so much that the product would solve the problem but that the problem existed at all.

Wu's point is relevant to contemporary ads for NIPT, which also tap into human ideals, press on anxieties, and draw on medical authority. But the same tactics are likely to be used in the selling of future biomedical interventions, particularly when enhancement is on the table. These, too, may promise "a new self, better than the old"; they may present options that people didn't know they wanted but find difficult to resist.

*   *   *

In ads for NIPT, women are depicted in ideal terms. The portrayal of people with testable conditions is something else again.

Unlike the consumers of the tests, who are portrayed as smiling, concerned, happy, and photogenic, people with Down syndrome and other testable disorders exist only as blurry, medicalized abstractions. The most common words used to describe them are *defects, risks, abnormalities*. None of the company websites show pictures of people with Down syndrome or any of the other conditions detected by NIPT. It isn't simply that genetically normal children are presented in a positive light, and that children with genetic disorders are presented in a negative light—though this, with token positive gestures, is true of the company advertising. It's that one group is presented as *real*—that is, as individuals, as an occasion for congratulations—and that the other is *abstract*, a list of problems, and a cause for concern. (That asymmetrical binary is distilled by the name *MaterniT21*. A person can be maternal; a cell has a trisomy.) One group is composed of individuals with happy stories, and the other group is rendered in terms of risk.

It is understandable, of course, that none of the companies in question feature smiling children with Down syndrome on their home pages. But that is precisely the point. Because the test's main selling point is that it provides peace of mind, the need to market the test creates pressure to blur and oversimplify. So there are conspicuous absences. Studying the websites, I never saw photos of people with Down syndrome or of any disabled people at all, including the mothers. Nor did I see the word "abortion" mentioned, though the insistence that the test can be done "as early as ten weeks" is a clear gesture to the possibility. As of 2018, for instance, Sequenom's account of Down syndrome offers an odd hybrid of risk and welcome:

> Down syndrome is a genetic disorder that results in mild to severe disabilities, and is caused by an abnormality in the number of chromosomes an individual inherits from their parents. Chromosomes are the structures that carry our genes, which are small sections of DNA that determine how we grow and develop.

"Mild to severe disabilities" is vague but technically true. However, for intellectual disability specifically, most children with Down syndrome fall into the mild to moderate range. The account continues:

> Women over the age of 35 are known to have a higher prevalence of children affected by Down syndrome. If you're expecting and worried that your baby may be born with Down syndrome, and would like more time to prepare for the birth of your child, tests are available to offer further insight into your pregnancy.

The websites are moving rapidly, and they have seen improvement over the years. But these improvements are ultimately superficial. They take place in the margins: a slightly less toxic description of Down syndrome; a slightly clearer acknowledgment that raising a child with a disability "can be rewarding"; a link to better information in another website. But the main impact is in the contrast between idealized images and risk-laden descriptions, and over the last five years, that pattern remains constant.

Where future children are concerned, our sense of risk itself has already been altered by the advent of prenatal diagnosis. In *Telling Genes*, her history of genetic counseling, Alexandra Minna Stern notes that with the arrival of amniocentesis in the 1970s, "prenatal diagnosis created a new landscape for genetic risk assessment and reproductive decision making." With the possibility of diagnostic certainty, the "thresholds of acceptable risk" were radically lowered. But amniocentesis is a technique, not an algorithm; unlike NIPT, amniocentesis is not the intellectual property of a single corporation, and so information about it came not from advertisements but from medical professionals.

Because NIPT is advertised online, the websites can structure information to the test's advantage. This is why the pictures of ideal families are important: set beside them, disability is not merely an abstract risk, a percentage, but a risk *to* something, to the combination of ideal children, motherhood, and family, to the vision of a good life. The test, then, protects the pregnant woman from risk and carries no risk itself. It becomes a means of guaranteeing the happy alternative. Though the competing companies vary in their approach, their images and rhetoric combine to express a eugenic logic: disability is defined as a problem, and science offers the solution. To see how this works, we need to examine the depiction of the test itself.

*   *   *

It may seem odd that a test labeled *noninvasive* begins by piercing the skin, but the label is key to a message that dominates the test's marketing: that it's both accurate and safer than invasive diagnostic tests. This, as we saw, was a leading talking point on the *Today* show segment where Sequenom's MaterniT21 PLUS was featured, and it has remained so ever since. From a rhetorical perspective, though, it's worth noting that the selling begins with the ordinary noun. Genetic counselor Robert Resta, writing at *The DNA Exchange* blog, notes "the subtly misleading implications of the name Non-Invasive Prenatal Screening":

> Sure, NIPS is non-invasive. But so is ultrasound, AFP [alpha-fetoprotein testing], HCG [human chorionic gonadotropin testing], etc. All of these screening tests are non-invasive and therefore do not carry a direct risk of fetal loss. NIPS is no different from the rest in that sense. It is superior to other screens in terms of having a very low first positive rate, high positive predictive value, and high sensitivity. But NIPS is still an alternative to other screening tests, not to amniocentesis or CVS.

Resta then notes the insistent comparison between amnio and NIPS on corporate websites, with amnio's chance of pregnancy loss repeatedly emphasized.

Resta's post was written in 2014. Since then, likely because of the efforts of Resta, genetic counselor Katie Stoll, and a few others, the websites have improved somewhat. The word "screen" is used more often; on one of its pages, Sequenom is careful to distinguish between diagnostic and screening tests, a different approach than the one taken on the *Today* show. But the key words—*safe, simple, accurate*—are still the ones in bold, and the risk from amnio is still noted. A web animation at natera.com, in flowing, pastel, faux-watercolor lines, walks us through the testing process: slim, contented woman hugging pregnant belly; vial of blood, with different-colored DNA fragments afloat in it; picture of computer screen showing the word NONINVASIVE; picture of woman with doctor; vial of blood, again, this time rising strangely from *behind* the woman's arm, then taking over the screen. The voice-over tells us, *no risk to your baby*. As she says it, her words spell out above the vial. Illumina, owner of the verifi test, emphasizes the downside of other tests: "Other types of prenatal screening and diagnostic tests may require more than one office visit, multiple blood draws, or carry a higher risk of false positive results. Diagnostic tests, such as CVS or amniocentesis, provide definite results for most chromosome conditions but have an associated risk of miscarriage." On the same page, a yellow headline announces, "NIPT Is Noninvasive to the Mother and Baby."

To respond to the advertising for NIPT is to respond to a corporate voice. It is anonymous, thoughtful, expert, synthetic, reassuring, incorporeal, and it is engineered to solve a complex rhetorical problem: how to make an uncertain test seem certain, a world of technical complexity seem simple, a new technology seem shiny-new and not scary-new, and persuasion sound like information. That voice is distilled in the bold print catchphrases and headlines associated with each test: *Find Clarity Early* and *Demand Clarity* (Ariosa's *Harmony*), *Highly Accurate, Comprehensive Results You Can Trust* (Natera's *Panorama*), *The Reassurance of Knowing* and *Empowering Informed Choices* (Illumina's *verifi*), and *Pioneering Science, Personalized Service*, and— my favorite—*Highly Accurate Answers to Important Questions* (Sequenom).

These phrases crystallize the ads' appeal: they link science words with emotion words, linking *accurate* to *trust*, *reassurance* to *knowing*, *pioneering science* to *personalized service*. By doing so, they imply the message beneath the message: that science and technology, embodied by the test, allow us to achieve a good (American) life. An earlier Sequenom campaign (circa 2014) made this explicit, with its trademarked *Quality of Science*, an unintentionally revealing phrase: as with its accompanying catchphrases—*The Science of Delivering Results Confidently, The Science of Results You Can Trust*—the phrase had less to do with science than a *feeling* of science, a quality. In one way, the catchphrases resemble the tests they advertise: their meaning is hedged, uncertain, indefinite, a matter of interpretation. *The Reassurance of Knowing*.

I've always been interested in tiny forms: aphorisms, captions, epigraphs, epitaphs, headlines, slogans, catchphrases, classified ads, haiku, tweets, punch lines. Tiny forms have two notable features: they depend on suggestion (because there's no room to expound or explain), and they therefore depend on context, on the beliefs around them. But their relation to context varies. Punch lines, for example, are closely bound to an immediate context: they need the joke's setup to make sense. In Japan, haiku are understood in terms of an elaborate set of seasonal motifs: culturally specific traditions of reading illuminate the form. Aphorisms are understood as vehicles for wisdom or advice; like haiku, they attain compression through image and metaphor.

The catchphrases attached to NIPT function by suggestion, but unlike haiku or aphorism, they are determinedly nonspecific. Like the endless commercial messages we marinate in daily, the pitches for NIPT trade in feeling, but their language is stripped of vividness and conceptual precision. This is partly because of the medium: a website allows specifics to be cited elsewhere and lets the imagistic function be displaced into actual images. The context is carefully shaped. But if the catchphrases are vague, it's also because of their audience: they're *meant* to be vague, so that the consumer can fit them to her life. They are as generic as the accompanying images of motherhood and family. They contain a nonactionable promise, like *results you can trust*; they establish a verbal space where the test, a hoped-for-future, and a prospective consumer's life can believably coincide.

In other words, the pitch takes the form of an implied story. The story is idealized enough to be appealing, generic enough for the consumer to become a character, and most persuasive when it appears to be blandly informative. Take, for example, Ariosa's explanation of the Harmony test:

> When you are pregnant, your blood contains fragments of your baby's DNA.
>
> Harmony Prenatal Test is a new type of test that analyzes DNA in a sample of your blood to predict the risk of Down syndrome (trisomy 21) and certain other genetic conditions.

In this brief, I-know-you-haven't-thought-about-science-lately explanation, what interests me most is the phrase *your baby's DNA*.

To say it's "your baby's DNA" that's tested is persuasive on multiple fronts: it humanizes the "baby" as one to protect. The future child becomes a character in the mother's story, and the pronoun *your* emphasizes the mother's protective role, lightly implying her responsibilities. It is *her* job to use the technology to protect her future child. And in a subtle way, the phrase *your baby's DNA* implies the technology's reliability. If, as in amnio, the "baby's DNA" is being sampled directly, isn't the result that much more certain? And yet the claim is demonstrably false: the DNA in question is not fetal, but placental. Because of a phenomenon called "confined placental mosaicism," placental and fetal DNA do not always match, a fact which may account for the error rate with NIPT. (Most companies have backed off the phrase "cell-free fetal DNA," retreating to "cell-free DNA," just as NIPD—for *diagnosis*—has given way to NIPT or NIPS, for *test* or *screen*. Over the short history of the technology, the pattern is that bold claims are made first and then only retracted under pressure.)

That error rate is greater than the catchphrases might imply. In recent years, investigative reporters have uncovered stories that belie the statistics. Beth Daley, an investigative reporter with whom I shared a panel at the NSGC conference in New Orleans, wrote this in the *Boston Globe*:

> a three-month examination by the New England Center for Investigative Reporting has found that companies are overselling the accuracy of their tests and doing little to educate expecting parents or their doctors about the significant risks of false alarms.
>
> Two recent industry-funded studies show that test results indicating a fetus is at high risk for a chromosomal condition can be a false alarm half of the time. And the rate of false alarms goes up the more rare the condition, such as Trisomy 13, which almost always causes death.
>
> Companies selling the most popular of these screens do not make it clear enough to patients and doctors that the results of their tests are not reliable enough to make a diagnosis.

The performance of NIPT depends on prevalence: the rarer the condition, the less accurate the test is. In the case of Down syndrome, which is correlated with age, the older you are, the better the test works. For a woman at 40, for instance, a positive result is strong, but short of diagnostic: it has a 93% chance of being true. If she's 35, the test is 79% accurate; put another way, one in five positive tests will be incorrect. At 30, the accuracy is 61%; at 25, the accuracy is 51%. This is not meaningless information, but it's also a coin flip: every other test will be a false positive, a fact inconsistent with words like *reassurance* and *clarity*.

In her article, Daley noted that in a study performed by Natera, a significant number of women aborted based on NIPT results alone: "22 women out of 356 who were told their fetuses were at high risk for some abnormality terminated the pregnancy without getting an invasive test to confirm the results." And in a likely testament to the power of marketing, one woman in a Stanford study "actually obtained a confirmatory test and was told the fetus was fine, but aborted anyway because of her faith in the screening company's accuracy claims." Athena Cherry, the professor of pathology who examined the fetal cells, said, "She felt it couldn't be wrong."

Studying the websites, I realized that it was the fine print that mattered most: not the boldface 99%, but the fine print 5%; not the boldface *noninvasive*, but the fine print *may require invasive testing for confirmation*. It was the dissonance between bold and fine print that bothered me, a dissonance familiar from ads for Viagra or Effexor a dozen other drugs: the visions of happy, telegenic, diverse-ish upper middle class lives (potency reawakened, functional mood restored), accompanied by a resonant, friendly recitation of terrifying side effects. The difference is that pharmaceutical ads are regulated, and ads for laboratory-developed tests are not, so in the drug ads, the dissonance (because mandated) is easier to notice.

We're familiar with the corporate poetry of drug names: Paxil (peace), Celebrex (Celebration + RX), Abilify. Shotgun weddings of Idea and Pill, Desire and Science. However, drugs treat the human while genes are where the human begins, and so, by the standards of what the genetic counselor Robert Resta calls Big Genoma, Big Pharma's names are small potatoes. Natera, Verifi: nature, truth. If Big Pharma's presiding god is Psychiatry, Big Genoma's is Philosophy. And yet NIPT is effectively being sold as a drug. What the ads promise is a feeling: reassurance, peace of mind, freedom from anxiety. What they do not ask is where the anxiety comes from or why it is that we feel particularly anxious about disability. Instead, they stoke that anxiety in order to sell the test.

*　*　*

Since their inception, the websites marketing NIPT have advanced two core arguments to prospective consumers. The first might be called the Goldilocks argument: Typically the test is positioned as *more* accurate than conventional screening and *less* invasive (and earlier) than amnio or CVS. It is the "just right" test. This argument depends on contrast, emphasizing the way NIPT is *different from* other prenatal tests. Confusingly, it's often combined with a Medical Authority Argument: at times, NIPT is framed as being *just like* other kinds of tests. When it serves the purposes of marketing—as, for example, when citing the recommendations of a medical organization *in general* or when noting that women often use prenatal diagnosis—the line between NIPT and other kinds of testing is blurred.

That framing of the test is part of a larger pitch. The ads present an idealized pregnancy, motherhood, and family and then present testable conditions in terms of risk: as a threat to that ideal. In this view, the test—itself idealized, in terms of accuracy—is presented as a simple, safe way to reach the desired ideal. That implied logic is distilled by an image on Sequenom's website, circa 2014. In it, a mom who could've stepped from a Talbots catalog, white, blonde, late thirties/early forties, is beaming and holding up an ultrasound of a recognizably babylike fetus. The tagline: *Better Results. Born of Better Science.*

There are two problems with the way persuasion tends to work in online ads for NIPT, and both are at odds with the way good medicine is practiced. The first problem is that the persuasion is covert. Although it appears to be open and informative, it is composed of subtle appeals, of highlighted risks and implied reassurances. As such, it differs significantly from the communication we would want from a medical professional, which should be open and neutral, not coded and biased. The second problem is the very existence of persuasion, which seeks to influence a woman's decision in a specific direction rather than helping her articulate her values and come to her own. That effort skates over a central decision, one that precedes which test to use, which is whether to test at all.

6   New Orleans

In mid-September of 2014, I flew to New Orleans to talk to a ballroom full of genetic counselors and industry representatives about the way NIPT is advertised to American consumers.

If I were an investigative reporter, like my copanelist Beth Daley, I could discuss the regulation of laboratory-developed tests or the stories of women whose lives were upended by inaccurate results. If I were a genetic counselor, like Katie Stoll—the chair of the panel, with whom I'd spent hours discussing the tests in the weeks before—I could speak directly to the clinical realities of genetic counseling, to the industry-sponsored science behind the industry-sponsored marketing, and to the statistical manipulations that make the tests seem more accurate than they are. Since I am a writer, I talked about persuasion, focusing on the web pages aimed at prospective mothers. I wanted to highlight the charged values just beneath the neutral-sounding copy, to counter the common mantra *simple*—"a simple blood draw"—with the human complexities at stake. I hoped to do, in a way, what targeted persuasion does, to cut out the middleman between me and my hearer. Genetic counselors occupy a key position in the prenatal testing process: they explain the tests, and the conditions tested for, to prospective parents. Because they are also the targets of industry marketing, I wanted to speak to them in a direct and unfiltered way.

"Genetic counselor," unlike "writer," is a new job description in human history. The jobs are surprisingly similar. Both involve a delicate, complicated act of communication, involving recondite information and the deepest human mysteries. Only one, though, depends on the ability to process gigabytes of data with proprietary algorithms. You would have thought that I'd have gotten to know a genetic counselor or two before Laura turned thirteen, but I only came to know one as a writer and friend, not a parent, and long after I had any questions about what Down syndrome might mean for my family. By then, I had begun explaining Down syndrome to medical professionals, instead of the other way around. Working with Katie, and traveling to New Orleans, was part of an effort to both teach and learn.

I'd become friends with Katie in the run-up to the conference. I'd gotten to know her through another friend, Alison Piepmeier, the feminist scholar and writer and mother to Maybelle, who has Down syndrome. Alison and I had come up with the idea for the panel a year before, but Alison's brain cancer had returned, and she'd had to drop out because surgery, then a brutal course of chemo and radiation, were about to start. Beth, the reporter, had agreed to step in.

Alison died less than two years later. Her friendship was one of the gifts that accompany Down syndrome: the broadening of my world because of the people I meet as a direct or indirect result of Laura. Alison's daughter Maybelle is a few years younger than Laura, and they have not yet met, but I hope they will one day. Alison was fierce in principle and ebullient in person—my superhero nickname for her was "The Friendly Firebrand"—and though she was keenly aware of the problems with the messaging around prenatal testing, she remained radically open to talking to anybody. She had attended the same conference the year before and reported on one conversation on her blog. Her use of bullet points was typical:

- All the genetic counselors are women. All the drug reps are men. In suits.
- I'm a huge advocate of abortion rights, but it was a little weird yesterday talking to an MD who performs abortions. Among other things, she said, "This one couple saw that tv show with the kid with mosaicism [*Life Goes On*, with Chris Burke, who *doesn't* have mosaic Down syndrome], and they said, 'Our baby might be nearly normal!' I said no, that's not realistic."
- I didn't let this doctor know that I have a child with Down syndrome because I wanted to hear her real, unfiltered thoughts. And wow, were they troubling. For instance, she was *shocked* that people might adopt a child with Down syndrome. "Maybe it's a psychological thing," she said. "They'll never have an empty nest."
- Believe it or not, she actually told me that all people with Down syndrome get Alzheimer's. First, this isn't true. Second, it's something I criticized in my talk on Wednesday: do we need to be talking about Alzheimer's when a child isn't even born yet?

Outside hospitals and clinic offices, genetic counselors and parents of children with Down syndrome don't really tend to hang out. It wasn't hostility on my part, just a sense of the unlikeliness of it: genetic counselors explain prenatal tests for Down syndrome, mostly to people who want to avoid it; I have a daughter with Down syndrome. Awkward! And yet, talking to Katie, I'd discovered a kindred spirit. She was detail-oriented and alert to things that don't quite add up, which, if you spend much time around the marketing for NIPT, or the industry-funded science behind it, means being alert almost all the time. In the way I chewed over corporate koans like "The Reassurance of Knowing" or "Quality of Science," she would look at sample sizes and unstated numbers and then ask Emperor-Has-No-Clothes questions. *Why is the oldest woman in the study sample nearly forty-nine? Why aren't test results broken out by age?*

Genetic counselors like Katie help patients understand their test results. They cover much more than prenatal testing, of course—they also discuss cancer risk, for example—but across the board, they are interpreters in every sense: the last translators in a series of computer-mediated translations, a series that begins with a blood draw and ends in words. And yet in this context, even a simple word like *positive* is a door into a maze. The genetic counselor, at her best, has to help a prospective mother navigate the maze, and that not only means offering impartial information on the test and on the condition, but guiding that mother toward a decision in line with her own values. Does a "positive" result for Trisomy 21 actually mean that the fetus has Down syndrome? What does it mean to have a disabled child, for me, for my family?

As Alison noted, most genetic counselors are women. Katie, my Virgil in the fluorescent, industry-dominated netherworld of clinical human prediction, had told me so, but it was obvious throughout the conference. The long escalators, filled between talks, were nearly Y chromosome–free. Male genetic counselors exist but are relatively uncommon. I did see one, but his cargo shorts and T-shirt suggested he had gone looking for a Hacky Sack convention and gotten lost.

* * *

From the north, the Ernest N. Morial Convention Center is low, angular, and gray, with the building's title emblazoned in orange block letters, as if on the spine of an enormous book. It was Wednesday afternoon, the day before my talk. I had time to register, meet Beth and Katie, and get the lay of the land.

Entering the convention center was like stepping through an airlock. The air dropped twenty degrees: suddenly the world was organized, the atmosphere controlled. Inside, the phrase *National Society of Genetic Coun-selors* had a second meaning: I had entered a different society, a different world. All the humans were sorted and tagged. The ordinary categorical display of the city outside the center, the visible distinctions between pan-handlers, businessmen, tourists, construction workers, became granular, formalized, precise. As outside, the body's facts—ethnicity, sex, age—were wrapped in layers of interpretation, but the layers, and the correlations, were more specific. Clothes became uniforms. Most of the conference attendees were white, and most of the people working at the convention center were black. A classified world.

I collected my shoulder bag, name tag, and registration packet. In the packet was a yellow index card printed with the names of nineteen Partici-pating Exhibitors, and a blank place to receive a stamp, like the handstamps Laura used to get at the end of each gymnastics class. If you got your whole card filled out with stamps—no blanks, no exceptions—you could enter it for a chance of winning an iPad mini, a GoPro camera, a Kindle. On most days I feel at least ankle-deep in irony, but in the Exhibit Hall of the Ernest J. Morial Convention Center, it was hard to keep my head above water. The companies there were selling predictive certainty through advanced technology—the algorithm, a defense against the randomness of reproductive chance—and they were doing it with a *raffle*. For little computers.

Near the registration booth, in a mammoth, city-block-long foyer which adjoined the cavernous Great Hall Prefunction Space, which in turn yielded to an exhibit hall measurable in hectares, was a self-service booth that reminded me of Lucy's five-cent psychiatry stand from *Peanuts*. On it, above a jar of safety pins, hung colored ribbons with gold letters that read *PRESENTER, EXHIBITOR, VENDOR*, and so on. Now and then you saw someone wearing two or three ribbons at once, like a decorated general or a child with a prize animal at the county fair. But ribbons or not, it was easy to tell who was who. I glanced at name tags, sketching a rough mental map in my head of the civilization I had decided to visit. Power, status, affiliation: all were coded in clothes and bearing, and made explicit in the dangling tag. There were the industry reps, mostly men striding along in groups of two or three, dressed in suits with bright ties or open collared shirts. There were genetic counselors, almost all women in their twenties and thirties, their tags declaring their industry and academic affiliations. There were future counselors, young women from the Brandeis or Sarah

Lawrence programs, sticking together in groups. They'd go on to work for an obstetrics practice somewhere, a hospital. Increasingly, they're likely to work for one of the big labs, which market and sell one of the competing tests, process the vials of blood that stream in by the thousands from all over the country, and offer genetic counseling services to the patients wanting to understand the results.

Beyond the Great Hall Prefunction Space was the exhibit hall. I'd met Katie and Beth, and we waited with a minor crowd for the doors to open: it felt like waiting for a not-great Black Friday sale or a band that hadn't caught on yet. Katie had friends to catch up with, and Beth and I wanted to get a first glimpse of the exhibits before our presentation the next day. When the uniformed guards swung the doors wide, allowing the registered and duly tagged conference participants to stream in, we walked into what I would later decide was a vision of the future. At the time, though, it seemed both surreal and familiar: the online world had come to life.

Company names shone, towering above me: *Natera* was printed on a sign in the shape of a ring, suspended on wires from the distant ceiling, rotating slowly. Familiar catchphrases were emblazoned everywhere on posters and brochures. One of the videos I'd studied online was cycling on a large flat screen display: in it, a middle-aged woman in red offers a hearty *Congratulations on your pregnancy!* Then her voice downshifts: *But you may have concerns.* In my presentation the next day, I called this the "congratulations-I'm-sorry moment." Soon afterward the passage was deleted from the video. In the crowd, I recognized a genetic counselor, live and three-dimensional, from the same video. I felt the disorientation of seeing a celebrity in real life, the weirdness of a screen existence conjured to dimensionality, a body scaled like your own, in the same space.

Usually, Theresa goes to conferences like these, and for years, she brought back free Post-its and penlights and other tchotchkes. In recent years, most medical conferences have cut way back on the free swag, but the world of NIPT and genetic testing is new, awash in money, and not all that regulated. This meant lots of free pens. I walked from booth to booth; the sales reps stood waiting, cheerful, relaxed, eager to chat, handing out bobbleheads, plastic zebras, foam sperm in blue, pink, and purple, Post-its, spiral notebooks, pens imprinted with URLs. There was a free massage station sponsored by Illumina. There were abundant and excellent hors d'oeuvres. Since the timing of my flights from the West Coast had left me ravenous,

I stacked up the Lilliputian pita triangles beside a squishy little flag of dips, red and green and white. I drained a Coke, feeling the day-long headache ebb away, replaced by something more like giddiness, with a touch of vertigo. I was jazzed to be there at last, even if the whole scene seemed a promissory note for a coming/here dystopia. I got free pens from informaSeq and the Boulder Abortion Clinic, the only exhibitor with an armed guard in attendance.

It was all, as the cliché goes, *surreal*, but at the same time it seemed deeply typical, only a supersaturated version of our text-dense world. The exhibit hall seemed a Times Square of the genetic future, every screen and surface imprinted. And everywhere I looked were images of the double helix, three- or four-rung ladders pinched at the ends, twisted into emblems, and woven into company names: the *logos* braided into each logo, the word that codes for flesh. All around me, on tote bags and banners and pamphlets, they floated free, like cell-free DNA itself: fragments to absorb our hopes, words to be read and understood.

*   *   *

Our talk was scheduled for eight in the morning. As a poet, my attendance expectations tend to be low. I've driven five hours to read to two people, with one being the bookstore owner and the other being a trapped patron too polite to leave. But in the end, hall attendance was pushing three hundred, mostly genetic counselors and industry representatives. A troop of Natera employees, all wearing bright red shirts for their 22q11 Deletion Fun Run, filed in and sat in a middle row, facing us directly. It seemed a deliberate gesture, sort of politely defiant, but it was hard to know for sure.

If you want to learn about America in the age of the genome, you could do worse than look at the Natera-sponsored Fun Run. So much seems typical of our peculiar time: our insistence on mixing fun with virtue; the byzantine ideas of "health" implied by a communal act of heart-healthy exertion, undertaken to raise money and "awareness" about a condition that is variable, untreatable by any known means, unameliorable by exercise, and in fact only preventable by prenatal test and termination; and the whole thing sponsored by a company that stands to profit from said prenatal test. From the prominence of 22q11 on Natera's website, not to mention the Fun Run, you might not know that the condition's prevalence is 1 in 4,000. Put another way, a pregnant woman—without taking any test at all—can already be 99.975% certain that her fetus is free of 22q.

But the label *22q11* also shows how rapidly our understanding of biology is being altered by computers. Until very recently, the conditions grouped under "22q11" were thought to be separate. But the detailed mapping of the Human Genome Project made it possible to locate the q11 deletion on chromosome 22. That mapping, in turn, was made possible by exponentially increasing processing power, which makes possible, among other things, smartphones, laptops, and NIPT. *Down syndrome* becomes *trisomy 21; DiGeorge syndrome, velocardiofacial syndrome,* and *conotruncal anomaly face syndrome* are aggregated and relabeled as the *22q11 deletion.* Our names for conditions are, increasingly, numbers, as if the algorithms were finding their self-portrait in us.

22q11 is what's called a "microdeletion"—one of the expanding panel of conditions for which NIPT can provide results. Its effect is widely variable: children can be severely to mildly affected; one genetic counselor I spoke with said she had had someone come in for another reason and only learned that he had the deletion because of the test. Natera had announced that they were raising money for the 22q11 foundation, and that they would match up to $30,000 in contributions, which—given their likely travel budget to New Orleans, the massive display in the exhibit hall, the cost of being a Silver Sponsor ($5,000), the logo-imprinted Post-it Notes Cubes ($3,500), and the full-color inside front cover ad on the Program Book ($2,000)—didn't seem all that much.

The marketing benefits of sponsoring a parent group are obvious. Partnering with parent and advocacy groups is a tried and true strategy of drug companies: the nonprofits are so poor, and the companies are so rich, that the money is hard to turn down. Partnering with parent groups blunts objections from all sides—pro-life activists, disability rights activists, and so on—and humanizes the company.

There are good business reasons for highlighting the new microdeletions. First, and most obvious, the ability to detect more conditions makes the test more appealing. Second, as Katie Stoll pointed out in a blog post that year, NIPT was not then recommended for women in "low-risk" categories, which meant that the test was mainly restricted to women over thirty-five; however, microdeletions like 22q11 aren't age dependent, so expanding NIPT to those conditions offered a back door into a lucrative market. And third, since no proprietary algorithm is markedly superior to the others, focusing on a single disorder sets a company apart, helping to define a brand.

* * *

On the second day of the conference, I boarded the St. Charles trolley and rattled toward the Garden District. The talk had gone well, but it was over. The exhibit hall had lost its enjoyable surreality; it was bleak and too bright, and the afternoon presentations were either incomprehensible (*Congenital Disorders of Glycosylation: Clinical and Genetic Variability*) or depressing (*Termination of Pregnancy for Indications of Genetic Disorder in Advanced Gestations*). The polished, thickly lacquered wood of the window trim reminded me of the older trams in Melbourne, where I'd lived for a year. Running my hand along the trim, I felt far from the shining, sunless world of the conference.

Visiting a new city, one skates like a bug on the surface tension of images, never breaking through. The night before, I'd walked through the French Quarter with friends I'd known online for years but only just met, taking in the sights that seemed equally real and emblematic: Bourbon Street at night as we tacked through tourist crowds; stolid bouncers in black T-shirts; a blues band visible through an open door, a white guy with a squinched face wailing pentatonically on a Strat; a waiter settling a linen napkin in my lap; wrought iron, accordions, Abita beer, beignets glistening in heaps of powdered sugar. The self-conscious catchphrases of a city marketing itself: *Nawlins*, *Big Easy*, *Yat*, the city presenting itself emblem first.

The tram moved heavily away from the central business district. I felt free. Behind me was the temperature-controlled convention center, the narrow streets, the tall, crammed-together buildings. I got off at Washington Street and walked toward the river. Through the entrance to Lafayette Cemetery #1, I could see stone surfaces weathered to shades of white and gray, as if the dead were shopping for paint colors: *New Moon*, *Vellum*, *Bone*. There was a small crowd milling at the entrance—a tour group, I assumed—but I later heard that the cemetery was being used to shoot a scene for a new episode of *NCIS: New Orleans*.

It seemed to me that in the convention center, I was taking notes on the future, while in the Garden District, with its weathered, above-ground graves, I was taking notes on the past. It did not seem possible that the two existed in the same city, or that either seemed part of the mild and humid present. The simultaneity strained the imagination: it seemed to me that the conventions already assumed in the convention center (that more information is always better, that it is okay to sell that information, that health should be as expansively defined as possible in order to choose certain kinds of children) did not square with the conventions illustrated by the cemetery, the honoring of the dead, the lifting of their remains above the possible floods; and that in another way, in a city below sea level, kept there only by the force of technology, by the defiance of nature, that the convention center was perfectly concordant with the city in which it found itself, and with the species that built the city.

As an attraction for walking tours and a set for NCIS, the cemetery's function overlapped with that of the convention center: it helped bring tourist dollars to the State of Louisiana and the City of New Orleans. But it did not exist for that purpose, and its existence suggested other ways of seeing the human than those expressed in the exhibit hall. Not the obsessive delineation of normalcy and health, not the focus on science and the future, not the celebration of technology, of profits and results. The windowless city of the cemetery, open to the weather and the changing light. It seemed the opposite of the world conjured in the exhibit hall, which seemed ultimately driven by the fear, even the denial, of death.

* * *

For a long time after the conference, a cube of Post-its sat on my desk, each peel-off page printed with a ghostly *Natera*. As the months went by, it shrank until it was gone. I was gutting and renovating our house then, jotting down lists of parts that would disappear behind drywall: Romex, PVC elbows, 2×4s and 2×6s, spools of cat 5e data cable. Elsewhere, across the country, genetic counselors were writing down grocery lists, phone numbers, notes to self. On all those discarded scraps, *Natera* was a faint watermark beneath the ephemera of each day, in the way our lives are watermarked with the trademarked names of corporations, the text beneath the text.

Though I'd gone to New Orleans to speak about NIPT, one of the prominent topics of the conference was "whole-exome sequencing": scanning all of the protein-coding regions of the genome. In theory, if we can predict the likelihood of Down syndrome, we can predict the likelihood of other conditions that are wholly and partly genetic, and given the current lack of regulation, it seems likely to me that these tests will be sold before they are much discussed. Deep wells of feeling, as the wedding and funeral industries teach us, are sources of profit, and the feelings attached to our children, our wishes for their health and happiness, are deepest of all. It will be difficult to resist testing for everything, even when the tests—as they are now—remain uncertain and oversold.

When I think of our rapidly increasing, fine-grained knowledge of human genetic variation, I imagine the pans of a giant balance. On one side is the gigantic and growing pile of genomic data, and on the other side is an equally gigantic and expanding idea of abnormality. One is the consequence of the other: faced with the overload of information, we invoke *abnormality*, a word capacious enough to hold everything, and everyone, considered different and undesirable. NIPT tests for a growing number of conditions; each is variable and different from the others; the predictive value of the test varies too, often performing worse with severer, and rarer, conditions. Perhaps, given the conceptual challenges of parsing the differences between disease and disability and variation, we will let the algorithms do the work for us, seeing as "abnormal" that which we can detect or claim to detect.

Flying home, I thought of two possible outcomes. One is that if we are all tested for our risk factors, we might return to the common sense that has, at least, worked for me: we *all* have a host of unknown risks, most of which are not treatable, or whose "treatment," like sensible diet and exercise, are things we might do anyway. So if we come to see everyone as having a list of "problems," of potentialities, then the line between "normal" and "abnormal" might erode or even dissolve. If we are all sick or pre-sick, then perhaps we will find common ground in frailty. Normality is a fiction, as is the idea of a life without risk; perhaps we will let these fictions go.

Or perhaps, finding the line between "normal" and "abnormal" untenable, we will trace a new line in a different place. Perhaps, weighed down by specific and uncertain fates we can do nothing about—a greater chance of liver cancer, of Alzheimer's, of addiction—we will try to alter the genes associated with those risks. Perhaps the list of abnormalities will be so long that we will have no "normal" to stand on anymore. In that event, I wonder whether we will be able to resist the pressure to enhance: not only to detect and select, but to rewrite what we will be.

7 Reading Synthia

In 2014, chemist Floyd Romesberg, of the Scripps Research Institute, synthesized a new pair of artificial nucleotides and got a cell to accept them as part of its genetic code. In metaphorical terms, he extended the alphabet of life.

To review, the DNA molecule is built from four nucleotides, or "letters": adenine (A), thymine (T), guanine (G) and cytosine (C). Each letter is one half of a pair—A always goes with T, and G with C—and each pair forms a single rung of the molecule's twisted ladder. Romesberg's team, after years of work, synthesized a third pair—X and Y—and inserted it successfully into the code of a bacterium, which then reproduced, maintaining its synthetic code. Life on earth depends on a four-letter code; Romesberg had invented a life form with six. In 2017, he updated the accomplishment, optimizing and stabilizing the cell. More important, he showed that the cell could express a novel protein. "We stored information, and now we retrieved it," Romesberg told *The MIT Technology Review*. "The next thing is to use it. We are going to do things no one else can."

Discussing his project with the *New York Times* in 2014, Romesberg used a metaphor: "If you have a language that has a certain number of letters," he said, "you want to add letters so you can write more words and tell more stories." In his TED Talk, he extended this metaphor, asking the audience to imagine a typewriter with only four keys. Wouldn't six keys be better? Couldn't you say more? The metaphor seems flawed to me. It may be that new nucleotides = new amino acids = new organisms, but it does not follow that new letters = new words = new stories. I know lots of writers, but none of us have been thinking, "If only there were thirty letters in the alphabet, then I could finish my novel, *The Story of Jimβθ!*"

Romesberg is careful to separate himself from wings-and-ultraviolet-vision transhumanism. His declared goals are squarely, soberly medical: a six-letter alphabet could code for a larger complement of possible amino acids, which could be assembled into proteins not found in nature, which might be useful as medicines, which Synthorx, the company Romesberg cofounded, hopes to develop for profit. And yet Romesberg's metaphor points to a tension between expansive and restrictive views of technology. On the one hand, to "tell more stories" can be glossed as "creating as many novel life-forms as possible." On the other hand, Romesberg defined those "stories" in familiar, clinical terms—curing disease, including cancer—and the cells, Romesberg noted, would remain obediently in the lab, dependent on a diet of artificial nucleotides to stay alive. Their meaning would be circumscribed, contained, their lives kept safely in vitro. Romesberg's rhetoric walks a familiar line between old and new, "natural" and "synthetic." In this, it mimics the application it serves, which splices old and new nucleotides, natural and artificial, together. But more significant is the fact that Romesberg uses metaphor at all: that he uses literary techniques to persuade and does so self-consciously, reflecting on the materials of meaning. He behaves as, is, a writer.

*   *   *

For high-profile scientists, the ones who speak to lay audiences, write popular books, and deliver TED Talks, metaphor is a key persuasive tool. The right metaphor can soothe fears, explain the recondite, and familiarize the unfamiliar. It is scary to say, "We want to create, not only new life, but a new *kind* of life, one fundamentally different from every single organism that has ever lived." It's less scary if creating new life-forms is just like telling stories. We associate stories with entertainment, meaning, and self-expression. Like the vaguely positive keywords anchoring ads for NIPT (*health, choice, empowerment*) or de-extinction (*revive, restore*), *story* shines a rosy, apolitical light on a technological development, familiarizing the new.

Beneath the metaphor of a story is another, one so ubiquitous as to go unnoticed: that DNA is a language, one which we "read," "write," and "edit." This is closely related to other information metaphors: that DNA is a code, or software. As Hallam Stevens explains in *Biotechnology and Society: An Introduction,* "[h]istorians have documented how 'information' and 'code' came to be powerful metaphors in molecular biology in the 1950s and 1960s." Stevens notes the pervasiveness of the metaphor—"It is hard to imagine it any other way"—but notes that it is not inevitable: "After all, the As, Gs, Ts, and Cs are not like English and Japanese. They are not really a language. Nor are they really a code…it is important to remember that information and code are *metaphors* rather than literal descriptions of how biology works on a molecular level." Historian of biology Lily Kay, writing in 1998, noted both the use and limits of the metaphor:

> There is no way of avoiding metaphors and analogies as heuristics in the production of knowledge, biological or otherwise, and the information discourse has been particularly powerful and productive. But metaphors have their limits, and analogies should not be confused with ontologies.

In thinking about DNA and disability and directed evolution, a first step toward clarity, toward disentangling the categories of life, language, and code, is to not take metaphors literally: to recognize when metaphor is being used, to explore its implications, and to recognize its limits. One limit of the DNA=story metaphor has to do with the way reading works. Reading is linear and one-dimensional. When we read a novel or poem or essay, we read one word at a time, in order, and even when we reread books, we understand them in terms of the sequence of the whole. It matters that chapter 1 comes first and chapter 23 comes last. But in the Book of Life, the linear paradigm doesn't hold. Nathaniel Comfort writes, "The old metaphor is not wrong; it is incomplete. In the new genome, lines of static code have become a three-dimensional tangle of vital string, constantly folding and rearranging itself, responsive to outside input."

If DNA is three-dimensional, in constant interaction with its cell, if it is constantly being "read" in different locations simultaneously, and some of the "reading" affects other "reading" in a dynamic, at-present-incalculably-complex set of interconnected feedback loops, then the "reading" looks very different from human "reading." So the metaphor has explanatory value, but its value is limited. (In a short article on genes and metaphor, John C. Avise acknowledges the practical use of seeing genomes as information. He also offers other metaphors—ecosystem, community, city—and argues that "metaphors can and should evolve to accommodate new findings.")

There are other differences. In books, the definition of every word is known; we do not know the function of every gene. Further, we expect books to be densely coherent. Even when they stretch and sprawl and wander, they still make sense—however "sense" is defined—at every level of form, from word to sentence to paragraph to chapter. They don't typically contain long, random character strings. But as Comfort writes, the book of life is a mess—"[I]f a genome is text, it is badly edited. Most DNA is gibberish." Lily Kay notes that the parts that *aren't* gibberish remain difficult to interpret:

> Once the complexities of DNA's context-dependence—genetic, cellular, organismic, and environmental contexts—are taken into account, pure genetic upward causation is an insufficient explanation. ... And when epigenetic networks are included in the dynamic processes linking genotype to phenotype (e.g., post-translation modifications, cell-cell communication, differentiation, and development), genetic messages might read more like poetry in all their exquisite biological nuances and rich polysemy.

Every metaphor breaks down somewhere. To *have* a story, and to *be* one, are not the same. George W. Bush can have a story, and so can Lassie, or a tapeworm. But none of these creatures *is* a story, something designed deliberately and in molecular detail by a single creator, written into existence, letter by letter, word by word. So when Romesberg says, "[Y]ou can write more words and tell more stories," his metaphor assumes (and normalizes) the idea that it is acceptable to design new creatures in the first place. It lessens the difference between the evolved and the designed: all are "stories." And it subtly shapes the hearer's sense of the technology in question, making it seem more powerful and more certain. The scientist, at his "typewriter," taps out a new "story." The metaphor emphasizes human intention and interpretive certainty, a message with a clear meaning, reliably reproduced.

Turning Romesberg's rhetorical *you* to a literal one, I would ask, If a new story is a new creature, then what stories do you want to tell? We have no cultural limit on stories, on their complexity or intricacy: will there be any limits on the stories told with the new letters, or on their ability to replicate, or on the ability of the designed creatures to interact with the evolved? Who will be our storytellers, and what will they believe?

* * *

In Romesberg's formulation, new organisms are new stories, a conceit made possible by the root metaphor that DNA is a language. But a project completed around the same time by the J. Craig Venter Institute took the metaphor a step further. In his book *Life at the Speed of Light*, Craig Venter himself—the brash, iconoclastic scientist and entrepreneur, and the institute's founder—described his project as the first "synthetic cell"; it was named *Mycoplasma mycoides* JCVI-syn1.0, but it acquired the nickname "Synthia."

You can tell a lot about a biotech application about the way it's named ("noninvasive," "de-extinction"), and Venter's new cell is no different: its formal name highlights the merging of the biological and digital. By hybridizing Linnaean and digital terminology, Venter indicates his view that we are at "the dawn of digital life," when life, because it can be translated into digital code, can "move at the speed of light." The name also denotes authorship and intellectual property: Venter's initials are inscribed in the organism's taxonomical name (JCVI, for J. Craig Venter Institute).

As many have pointed out, Venter did not synthesize an entire cell. Instead, his team began with the genome sequence of one bacterial species (*M. mycoides*), altered the sequence on computer, built it from scratch, and implanted it in the cells of a different bacterial species; the assembled genome then took over the new cells. The synthetic genome, over a million base pairs long, was assembled from pieces ordered from a DNA synthesis company, which took Venter's digitally composed sequence—the string of bases, or "letters"—and then chemically synthesized it and delivered it in short, overlapping stretches called oligonucleotides. Venter's lab painstakingly stitched these together into larger pieces, which were themselves stitched together into a full genome: a synthetic chromosome, which was then transplanted into a cell. The project took fifteen years. Venter emphasizes the precision required in the experiment: the transplantation failed repeatedly because of a single typo, a single misplaced letter in a key gene. When the transplantation finally succeeded, the DNA at the cell's heart had been human designed and human assembled, but the cell divided and reproduced as if it were natural. As *The Guardian* reported, the cell "paves the way for designer organisms that are built rather than evolved."

Announcing the cell's completion, Venter demonstrated an instinct for publicity, as *The New York Times* reported:

> At a press conference Thursday, Dr. Venter described the converted cell as "the first self-replicating species we've had on the planet whose parent is a computer."
> "This is a philosophical advance as much as a technical advance," he said, suggesting that the "synthetic cell" raised new questions about the nature of life.

In the same article, Nicholas Wade reported the misgivings of leading scientists who found Venter's technical achievement remarkable, his hype distasteful. Leroy Hood used the word "glitzy." Nobel laureate David Baltimore granted the technical achievement, but added, "To my mind Craig has somewhat overplayed the importance of this....He has not created life, only mimicked it." Gerald Joyce similarly noted the "power" of designing a genome letter by letter, but rejected the idea that the cell was "a new life form": "Of course that's not right—its ancestor is a biological life form." The public rivalries of scientific frenemies are a popcorn-worthy combination of *Mean Girls* and *Pacific Rim*, but beyond the gossip, the arguments are as rhetorical as they are scientific: how should scientists represent their work to the public? *To my mind Craig has somewhat overplayed the importance of this.*

In the *Science* paper unveiling the project, Venter is relatively restrained, but in his press conferences and in his book, his claims lie somewhere between science, philosophy, literature, and guru-like prophecy. Depending on the audience, the same synthetic cell is communicated in radically different ways. This rhetorical divide is characteristic of new biotechnologies—think, for example, of the difference between an ad for NIPT and a consent form signed by a patient—but is also traditional. Like many of his scientific forebears, James Watson in particular, Venter is understated in scientific publications and hyperbolic before the press.

In Venter's case, the hyperbole takes the form of metaphor. In *Life at the Speed of Light*, Venter's description of his synthetic organism is exuberantly synthetic, splicing together elements of life, writing, publication, software, and the Internet: "We were ecstatic when the cells booted up....It's a living species now, part of our planet's inventory of life." Throughout the book, Venter treats metaphor like an engineer stress-testing a metal, pushing it to the point of failure. His point is that the metaphor is not metaphorical:

> [DNA] is *in fact* used to program every organism on the planet with the help of molecular robots. [Italics mine.] ...
>
> All living cells run on DNA software, which directs hundreds to thousands of protein robots. ...
>
> Digital computers designed by DNA machines (humans) are now used to read the coded instructions in DNA, to analyze them and to write them in such a way as to create new kinds of DNA machines (synthetic life).

To drive his point home, Venter encoded messages in Synthia's genome. These, described as "watermarks," distinguished the creature as synthetic. Venter used a code, with triplets of DNA letters equivalent to letters of the alphabet, to spell out messages, including the names of contributors to the

Science paper announcing Synthia's existence. Also included were three quotations, in all caps (the code didn't include lowercase): one from James Joyce's *Portrait of the Artist as a Young Man* (TO LIVE, TO ERR, TO FALL, TO TRIUMPH, TO RECREATE LIFE OUT OF LIFE); a saying attributed to J. Robert Oppenheimer's teacher, SEE THINGS NOT AS THEY ARE, BUT AS THEY MIGHT BE; and WHAT I CANNOT BUILD, I CANNOT UNDERSTAND, a misquote from the physicist Richard Feynman. (The original: "What I cannot create, I do not understand.") These were initially presented as a puzzle to solve: also encoded in the genome was an e-mail address, so the DNA machines (human) who'd figured out what the DNA machine (Synthia) was saying could contact the DNA machines at the J. Craig Venter Institute and let them know. Like Romesberg's "stories" written in six-letter DNA, like Church and Brand's woolly mammoth, Synthia is conceived as a kind of message, but it takes that vision to a literal extreme.

To me, Synthia is an elaborate, clever instance of biological wordplay—more sudoku than poetry, but suggestive nonetheless, a puzzle that remains mysterious even on decoding. But according to Venter, Synthia's meaning is clear. It—she?—has two lessons to impart. First, it disproves "vitalism," the idea that something nonmaterial—spirit, a "life force"—is necessary for life to exist. Life is material. And second, Synthia proves that life is information. Venter stresses the point, saying that "[t]hese experiments left no doubt that life is an information system." His work offers "the proof…that DNA is the software of life."

Mycoplasma mycoides JCVI-syn1.0 is a quasi-literary text, inscribed in a cell. For that reason alone, it seems to me that its potential interpretations are more varied, more uncertain, and more interesting than the ones advanced by its author. Among Synthia's all-caps proverbs is the declaration WHAT I CANNOT BUILD, I CANNOT UNDERSTAND, which implies that its builders best understand its meaning; however, if the history of literature teaches us anything, it's that the author is the last person you should turn to when seeking the meaning of a work. What an inventor wants something to mean matters less than what the world chooses to make of it.

* * *

If Synthia were just a really short book, no one would bother with it. It's a cereal box version of Bartlett's Familiar Quotations: three quotes, a list of names, and an e-mail address. It wasn't written by the author whose initials enclose the whole, and one of the quotes is incorrectly transcribed. Worse, it performs the remarkable feat of making James Joyce sound like a bad motivational speaker: TO LIVE, TO ERR, TO FALL, TO TRIUMPH, TO RECREATE LIFE OUT OF LIFE! And yet a closer look is rewarding, because the more you look, the more the cell's meanings splinter into uncertainty, beginning with its name.

Synthia, I'd thought at first, was a clever bit of wordplay on Venter's part, the cell's name expressing its nature: *syn* for *synthetic biology*; the letter *S* substituted for *C*, suggesting the editing process by which words were encoded in its sequence. It turns out that the name was coined back in 2007 by the ETC Group, a Canadian civil society group fiercely opposed to the project. The name was intended as mocking, like "Obamacare," but the effort backfired: it was catchy, so it stuck, and soon it became a handy generic name, if only because "*Mycoplasma mycoides* JCVI-syn1.0" does not exactly roll off the tongue.

If the mockery misfired, it may be because *Synthia* fits easily into the rhetoric of invented life, the specific kind of whimsy of those who, playing God or not, enjoy playing with words: *Dolly*, the cloned sheep (named after Dolly Parton, because the sheep was cloned from a mammary cell); *cc*, the cloned cat; Hercules, the genetically engineered, supermuscular beagle; "*Eau d'E. coli*," a variant of *E. coli* engineered to smell good. In books like Venter's *Life at the Speed of Light* and George Church's *Regenesis*, the light-hearted, catchy names sit oddly beside the grand claims about life, science, and the future. It is as if someone had stuck a limerick into the *Odyssey*.

This dissonance points less to science than to a saturated media environment, in which extremity and novelty are rewarded. The two rhetorical registers of biotech futures—stentorian announcements of a New Epoch and catchy names for new animals—are simply two forms of novelty, two ways to distract an audience from distraction. You can do that with entertainment or wisdom, a joke or a truth, a witty slogan or a sonorous prophecy. *To delight and instruct*, in the age of social media: our new pitches map onto an old poetic goal, made more urgent by the sheer quantity of information we have to sort through.

* * *

What is the best way to read a cell? Synthia's declarations tend more toward prophecy than wit, but even ignoring the substance of its all-caps wisdom, the very fact of quotation is suggestive. Plucking sentences from context mirrors the project itself, which tears genes from previous contexts and installs them in new ones, both digital and biological. And who is quoted matters as much as what is said: pointing to Oppenheimer, Feynman, and Joyce elevates the idea of the iconoclastic genius, clearly implying that Venter belongs in their company. By implication, it is Venter who has TRIUMPH[ED] by RECREAT[ING] LIFE OUT OF LIFE, who SEE[S] THINGS NOT AS THEY ARE, BUT AS THEY MIGHT BE. And with the quotation WHAT I CANNOT BUILD, I CANNOT UNDERSTAND, Venter lays claim to a superior understanding of life—and causes the built object, the living cell, to ventriloquize the claim.

It is an odd thing for a cell to say *I*, odder still when that cell's existence is a team effort. But Synthia hovers between individual and group achievement. Venter's initials occupy the taxonomic name, but as part of an institute; the names of his coauthors are inscribed in the "watermarks"; the cell, with Joyce, elevates an individual ideal of achievement, but at the same time distributes credit, albeit less prominently, to the group. As a digital creature, its uncertainties reflect a digital age when the author is in decline; when old ideas of intellectual property battle with new ones; when everyone on social media is writer, reader, and publisher at once; and when so much of what we share is curated, appropriated, and snipped from one context and repurposed for another. The quotations illustrate this tendency, but their real function is to identify the cell *as* synthetic and not natural. The quotes are claims that stake a claim. Of course, there's no end of irony in using appropriated text to establish intellectual property rights, especially when one sentence is a misquote. A string of random characters would have been simpler.

To me, the altered quotation from Feynman is endlessly suggestive. WHAT I CANNOT BUILD, I CANNOT UNDERSTAND can be read at face value, as a declaration of the synthetic biology principle that life must be constructed to be understood, and yet, as a misquote, it reads as a flaw in construction. Human achievement is undercut by human error. Given Venter's insistence on the precision of his genome editing, the proofreading error is especially ironic. At the same time, it reveals which kinds of precision matter. DNA may be, in Venter's accounting, an information system, but some forms of information are clearly more important than others. Bacterial genomes are proofread down to the base pair; human sentences—eh, close enough.

Implied in Synthia's text, and made explicit by *Life at the Speed of Light*, is an idea about which kinds of knowledge matter most. There's a paradox in books like these: scientists take on the role of artists, but the arts are distinctly secondary. Even as the scientist is portrayed as a storyteller with an automated sequencer, a painter with a palette of nucleotides, the arts come off as a sort of third wheel of civilization. They aren't ways of understanding the world, loci of transcendent shared loneliness across time, the set of practices every culture has, and without which life would have no point; they're just a source of iconic explanatory examples. Whatever is cited tends to be famous and big. Venter notes Joyce; George Church, in *Regenesis*, explains that he wants to make new genomes synthetically, not simply copy old ones, because "[p]hotographing the Mona Lisa is not as impressive as creating it in the first place." In this figure, art is more decorative than structural. It is the trim in the house of science, the false columns in front, substantial looking but not load bearing.

In both *Regenesis* and *Life at the Speed of Light*, this attitude to art is rooted in an engineer's approach to the world. What matters is doing something, making something useful. This idea resonates with me, but not when it's used to demote other human endeavors that are necessary and useful in their own ways. George Church, for example, defines synthetic biology in opposition to something "self-indulgent":

> Synthetic biology is mostly about developing and applying basic engineering principles—the practical matters that help transform something academic, ivory-towerish, pure, and sometimes self-indulgent or abstract into something that has an impact on society and possibly even transforms it.

Venter, too, defines himself as a problem solver, setting himself in opposition to a fictional ivory tower:

> Richard Feynman issued a famous warning about the dangers of attempting to define anything with total precision: "We get into that paralysis of thought that comes to philosophers. ... one saying to the other: 'You don't know what you are talking about!' The second one says: 'What do you mean by "talking"? What do you mean by "you"? What do you mean by "know"?'"

Like Church, Venter—though he touted Synthia as "a philosophical advance as much as a technical advance"—implies a clear contrast between practical doers and philosophical yappers. And yet: isn't disproving vitalism, the point of Synthia, kind of...philosophical? Doesn't transcribing a message in a cell raise questions about language? And don't Venter's own metaphors, like saying that people are *DNA machines* or saying *the cells booted up*, raise questions about what it means to talk, to know, to be a "you"—about, in other words, language, knowledge, and people? These questions are *raised* by transformative biotechnologies. That they can be pursued to dead ends is no reflection on the questions themselves: any line of inquiry, in any field, can be sterile and pointless. The key is to consider the questions in a fruitful way. That begins with questions of power: who gets to speak, who is considered authoritative, and who is spoken about.

These are ancient questions, but the digital age renews them. The existence of a "programmable cell" blurs life and nonlife, organisms and messages. Language like *the dawn of digital life* or *E. Cryptor* or *H. sapiens 2.0* celebrates the blur, playing with it in words, but beneath the play is a message of control, the ability to build a cell to order, to make it serve a task. Both meanings, it seems to me, are evident in the word *watermark*, which embodies the durable and ephemeral: it could stand for life (its code enduring, its forms changing) or the Internet, where the folk belief that data live forever is belied by the fact that data tend to disappear, either drowned by the sheer wash of new data or simply lost. But *watermark* is a printer's word, not a poet's. A sign of ownership. A mark of intellectual property, inscribed in fluid life.

It took a scientist (Theresa) to point out another irony to me: the very fact that cells change as they evolve means that the watermarks will change. Left to their own devices, Synthia's descendants will evolve. Since the "watermarks" are embedded in noncoding regions, the cell has no need to preserve them. Therefore the list of authors, the quotations, and the e-mail address to contact will begin to degrade. They will be no more permanent than marks in stone; they will weather from the inside out, the author's intentions fading, letter by letter. As an organic book, one that can reproduce independently, Synthia is self-publishing, but it is self-revising too.

As a living creature, Synthia is a chimera, an engineered blend of two species. But as a living book, it is a chimera of minor forms, a vanity-press amalgam of title, aphorisms, contributors' notes, and copyright page. These, Russian-doll style, are all enclosed in the "watermarks," which are in turn enclosed in a brainteaser: the entirety of Synthia's legible text was presented as a puzzle for smart, science-oriented people to solve. It is, in other words, an intelligence test, but it emphasizes one kind of intelligence, selecting for those who—like Synthia's inventors—have a problem-solving mind-set and a brain for code, and who are digitally savvy and connected enough to contact the J. Craig Venter Institute with their solution. This does not describe most of the humans in the world, let alone those whose cognitive differences are regularly put forward as the targets of biotechnology. It is technology that divides, not technology that embraces. I prefer a different view of technology and a different voice. A voice that is open and questioning, and that begins and ends with people and thinks about how tools might fit, rather than beginning with the tool and assuming that people will find a place.

But the brainteaser aspect of Synthia also returns me to the phrase *feeble-minded*, which replaced *idiot* as the term of art at the beginning of the mainline eugenic era. The phrase implies that a good brain is a strong one, and that thinking is a force applied to the world. This view leaves out other ways of being brainy in the world, other forms of discernment, comprehension, and presence. So many of our most meaningful experiences of thinking leave nothing in their wake. You don't have to make something of a moment for the moment to be valuable. If the ability to hear another, to be with another without judgment, and to sit in the moment peacefully matter, then the wall between people with and without intellectual disabilities begins to erode.

*   *   *

At the beginning of *Life at the Speed of Light*, Craig Venter writes that "humankind is about to enter a new phase of evolution." At the book's end, he offers a vision of life transcending time and space:

> We are moving toward a borderless world in which electrons and electromagnetic waves will carry digitized information here, there, and everywhere. Borne upon those waves of information, life will move at the speed of light.

Perhaps we are approaching a "borderless world," whatever that means. Until then, some will move more easily across borders than others. As science fiction author William Gibson famously said, *The future is already here, it's just not evenly distributed*: the wealthiest today already move as if in a borderless world. Privilege equals weightlessness, and to have money is to be like money, able to slip from one state to another. To be free, in the Information Age, is to resemble information.

Our rhetoric is the dry run for the future: what may become our bodies is first performed in words. *E-life, biological teleportation, "the first self-replicating species that we've had on the planet whose parent is a computer"*: these are short stories condensed into phrases, teasers for next year's blockbusters. Catchphrases like these figure life as information. As such, they are quietly persuasive, rendering the reality more imaginable and hence more possible. But that act of imagination reaches beyond words alone: it's also on our screens, where the same idea that (literally) animates Synthia—the merging of digital and organic—is the premise of one entertainment after another. That this is so shows that we are, on some level, already conscious of the radical changes afoot, and that we want to know what they mean for us. We are streaming, thinking, watching.

Our entertainments, however, are less celebratory. On HBO's *Westworld*, wealthy humans pay enormous sums to live out fantasies of sex and violence among organic, programmable cyborgs. In episodes of the dystopian BBC series *Black Mirror* (the title refers to a smartphone screen), the digital lives in people, and people live in the digital. Virtual doubles of real people inhabit, or are trapped in, simulations: a *Star Trek*-like multiplayer game, an Eighties-themed digital afterlife. Networked devices are implanted in brains and eyes, allowing recording, playback, monitoring, and control. The movie *Ex Machina* features a seductive, sentient robot, but it's equally about the way the digital pervades our lives already. These entertainments are black mirrors, screens and windows at once, showing us our world as it shows us ourselves. Their subject is less the technology to come than the technology around us now.

Shows like these, no less than the writings of scientists, are a part of the ongoing conversation about where our biotechnology is taking us. Their popularity alone suggests that, on some level, we're thinking about what people will be. But in the last decade, one movie in particular fuses the themes of digital information, human enhancement, and disability.

The Amazing Spider-Man (2012) rebooted the franchise that began in 2002, but with a twist: the genetic enhancement of human beings is central to the plot. Just as in the first series, the high school student Peter Parker, bitten by a genetically modified spider, develops superpowers. This time, though, the spider comes from a corporation whose business model depends on "cross-species genetics," and Peter's nemesis—the scientist Curtis Connors, once the research partner of Peter's deceased father—is a man with one arm, who not only longs to generate a working limb, but dreams of a world where disability and disease are erased.

For Dr. Connors, pressure from his corporate employer pushes these dreams across a moral border. Oscorp's namesake CEO—Norman Osborn, whom we never see—needs the technology to extend his life. Facing the loss of his job and funding, Connors injects himself to prove his research can work in humans. He collapses, and in a gorgeously creepy scene, awakens to his new self: Peeling back a dry, plantlike husk, he discovers a new arm both fetal and disproportionately large, the veins visible through translucent skin. In the logic of the fantastic, from *Dr. Jekyll and Mr. Hyde* to *The Fly* to *The Amazing Spider-Man*, Faustian choices lead to swift transformations. The hand soon becomes a giant claw, and Connors acquires a tail, reptilian eyes, and green pebbled skin to go with his resonant English accent. Before long, he has torn his way out the back of a taxi and is throwing cars off a bridge.

The special effects are good, but as the plot shows, the movie's focus on genetic modification is more than incidental. When we first meet Dr. Connors, his research is stalled. What he needs is a key equation: the "decay rate algorithm," which his former partner, Peter's father, invented shortly before disappearing. (In the movie's opening scene, the young Peter discovers his father's ransacked office. As we come to learn, his parents were later killed by Oscorp.) When Peter discovers his father's briefcase in the basement and then finds the algorithm inside, he's discovering a literal inheritance, one that soon shapes his biological identity. Driven by a desperate wish to know more about his father, Peter goes to Oscorp, snoops around in a Restricted Area, and is bitten by a spider. The algorithm is key to Connors's transformation too: Peter gives it to Connors, thereby allowing Connors's research to finally succeed.

Making the algorithm central is a canny plot point, highlighting the convergence of biology and computing. What results is more than a tail or an ability to swoop through city streets: both Peter and Dr. Connors, because their bodies are partly created by technology, are set outside the ordinary links of human generation. Their relations, to each other, to the species, are warped and almost broken. In a weird way, they become brothers: both are altered by the same code. But then, Connors and Peter's father are brothers too: one Cain, the other Abel, the evil one killing the good. In yet another view, Connors and Peter are both children of Norman Osborn, the absent corporate dad. Or they are corporate products themselves. Or they are their own parents: Connors almost literally gives birth to his new self. (There's no genetic determinism here: their stories depend not only on the genes themselves, but also on the characters' choices, on what they make of their alterations.)

Interestingly, Peter's father didn't destroy the algorithm; he hid it for his son to find. This act suggests that the technology can be used wisely, but only by the right person, by the one either noble or smart enough to recognize it. The sword in the stone is now an equation.

It's not long before Connors has discarded his dreams of healthier humans, deciding to focus instead on making everyone else into lizards too. Even though this sort of thing is a standard career choice for mad scientists, *The Amazing Spider-Man* pursues the implications further than most. Late in the movie, Peter comes across a video recording in which Connors explicitly rejects one vision of biomedical science (healing disability and disease) and embraces another (improving the species itself). That this is more than boilerplate bad-guy talk, that questions of science are being deliberately raised, is clear from an interview with the producers. Discussing the Oscorp Tower, a fanglike sliver inserted via computer-generated imagery into midtown Manhattan, producer Avi Arad says, "It was more of a symbolic idea of the world of science." He continues, "It's a place where his father worked. It's a place where Connors works. And most importantly, it's a place where Peter aspires to be. One, because his father worked there and two, it stands for the ultimate advancement of science and biotechnology."

Yet we should not mistake *The Amazing Spider-Man* for a radical or comprehensive critique. Despite its clever plot, the movie expresses the "technology is neutral" point of view that's standard for the genre: the technology of enhancement isn't good or evil per se, the user is, and the technology merely expresses the user's heart. That dichotomy between good and evil is rendered in visual terms. *The Amazing Spider-Man* has two instances of "cross-species genetics," but only one becomes grotesque. Unlike Dr. Connors, Peter Parker still looks human: he has superstrength, better vision, better reflexes, and the ability to stick to walls and ceilings. He doesn't grow extra eyes or limbs, and his love life improves significantly, something which is difficult to imagine for Dr. Connors, at least in the human realm.

The same pattern holds in other movies of the genre. In *Captain America: The First Avenger*, the technology is neither mechanical nor genetic—it's a kitschy blend of colored liquids and high voltage—but it too reflects the heart of the man it is applied to, a point made explicitly by the scientist at the controls. Steve Rogers, the hundred-pound weakling with a good heart, becomes Captain America, while the Nazi Johann Schmidt becomes an even more evil Nazi: Red Skull, who, like Dr. Connors, is physically grotesque. In *Iron Man*, the weapons manufacturer Tony Stark invents an armored supersuit, turning himself into a weapon—but for good. His nemesis, the corporate master Obadiah Stane, uses the same kind of supersuit for evil, but Stane's suit is Goliath-like, monstrous and imposing.

What sets *The Amazing Spider-Man* apart, though, is the centrality of ability. The binary of Parker and Connors is familiar—hero and villain, good and evil, normal and grotesque. (In a visual medium, ugliness is an easy and familiar shorthand for evil.) But the more significant split has to do with the way they come by their anomalous bodies: Parker's occurs accidentally, an unwanted destiny, but he comes to embrace it; Connors's is deliberate, something he chooses as a reaction to an unwanted disability. This plot point fits with the argument made by Dorothy Nelkin and M. Susan Lindee, the authors of *The DNA Mystique*: in pop culture, DNA has a quasi-sacred status; tampering is taboo, hence punished. Along these lines, Peter Parker is innocent because his alteration is accidental, while Dr. Connors is hubristic, both Dr. Frankenstein and monster.

*   *   *

If the movies suffer from double vision, it's because we do too. We have conflicted feelings about technology, its effects on people, and the companies that provide it; our entertainments mirror that conflict. Technology is scary—dehumanizing, intimidating, world threatening—and, in every sense, awesome: superstrength, glittering 3-D simulations, blue ray guns in 1945. Human enhancement is playing God (Red Skull, *Iron Man*'s Obadiah Stane, Dr. Connors) or discovering your true humanity (Peter Parker, Tony Stark, Steve Rogers). Large corporations are faceless, powerful, and nefarious, but at the same time sleek, luxurious, and impressive. There are many practical reasons to have a corporate bad guy—for a start, no one really gets offended, since corporations aren't actually people—but at the same time, the movies themselves, sophisticated, effects laden, are celebrations of the corporate application of technology: the power to make a lizard talk, a human being fly, a new skyscraper appear in midtown Manhattan. In the movies, if not in our bodies, the human-encoded creatures interact with the regular ones, the enhanced with the unenhanced.

The Amazing Spider-Man, more than most, leans toward a cautionary note. Its plot embodies the law of unintended consequences, the dangers of intermingling profit and research, and—in the person of Peter's father— a principled decision *not* to follow up on a discovery. But this cautionary approach, as convention and the market dictate, is swaddled in a plot about heroes, in which the fate of crowds depends on the heart of a single man. Therefore, the plot's very structure offers reassurance: it tells us that not only will justice prevail, but that individuals have agency, that what they do matters, and that a good heart wins in the end. But it is not a single evil user, but billions of ordinary consumers, following ordinary desires— driving cars, wanting healthy babies, going to the movies—that actually shape our world. Whether a movie like *The Amazing Spider-Man* trivializes its message with entertainment, or even normalizes the enhancements it seems to warn against, is unclear, but it is clear that no single hero, even an enhanced one, is likely to save us.

8   Dismissive Narratives

From thalidomide to global warming, short-term risk-benefit analyses have led us down paths of irretrievable harm. What we have failed to imagine has been excused as "inconsequential," swept under the rug as "side effects," lessons learned rather than lives ruined. The post-War aversion to eugenics—the understanding that despite great variability from one human to another, no one life is worth more than another—has eroded. Never have we more needed thoughtful, unrushed and thoroughly democratic models of transparency, public discussion, and distributive justice.

—Patricia Williams

In 2015, Harvard psychologist Steven Pinker published an op-ed in the *Boston Globe*, extolling the possibilities of CRISPR-Cas9. Though he acknowledges that biomedical research is "incremental" and "slow," his article leaps ahead, nonincrementally, to extravagant promises: "Physical suffering and early death have long been considered an ineluctable part of the human condition. But human ingenuity is changing that apparent fate."

Skating over the complications, real-world problems, and ethical conundrums of human-focused biotechnology, Pinker instead opts for a crudely personal pitch:

Have you had a friend or relative who died prematurely or endured years of suffering from a physical or psychiatric disease, such as cancer, heart disease, Alzheimer's, Huntington's, Parkinson's, or schizophrenia? Of course you have: the cost of disease is felt by every living human. ...

Just imagine how much happier you would be if a prematurely deceased loved one were alive, or a debilitated one were vigorous—and multiply that good by several billion, in perpetuity.

Pinker appeals to the reader's emotions: *Just imagine how much happier you would be.* But few responsible scientists would imply that cures for heart disease, cancer, or Alzheimer's are immediately forthcoming; that bioethicists, as opposed to the complexities of biology, are the primary obstacles to those cures; or that said cures could be easily scaled up to a population level. (They might be too expensive to produce; they might only work in certain subcategories of disease; political or economic barriers might deny access. On this last point, Pinker evinces a rising-tide-lifts-all-boats serenity: "As the treatments get cheaper and poor countries get richer, these gains will spread.")

So far, so inflated. But the backlash against Pinker's piece had less to do with his hyping of CRISPR and more to do with his account of the conversation. Pinker attacked bioethicists as obstacles to progress and health. Though he criticized them for favoring excessive regulation, his real venom was saved for what he saw as, in effect, literary sins: invoking "science fiction dystopias," raising "perverse analogies with nuclear weapons and Nazi atrocities," and "sowing panic about speculative harms in the distant future." For this simple problem, Pinker had a one-sentence solution: "Get out of the way."

As a tenured academic dismissing an entire academic field, Pinker resembled a politician who, in a bid for status within a system, pretends to be outside it. It's a maverick's pose: say things shocking enough to go viral, but not shocking enough to disqualify. Like other faux outsiders, Pinker takes a simple approach: sketch a simple, moral narrative, with an obvious problem and an obvious solution; populate it with good guys (researchers who heal) and bad (bioethicists who obstruct); and inject a crude emotional appeal ("just imagine how much happier you would be"). The Internet, duly infected and feverish, keeps you in the news, and the outrage only confirms your outsider status.

Pinker's broad dismissal of bioethics was rightly criticized. Far less discussed was his implied dismissal of people with disabilities. Throughout the article, he assumes uncomplicated equivalences between "disability," "suffering," "disease," and "death." In a follow-up interview on the blog of stem-cell biologist Paul Knoepfler, Pinker stated his view more baldly:

> You'd think it would be an obvious ethical principle that life is better than death, health is better than disease, and vigor is better than disability. But, astonishingly, so-called bioethicists have repeatedly denied these truisms, either explicitly (in

the case of the country's former bioethicist-in-chief, Leon Kass, who argued that the desire to extend life is a sign of shallowness and immaturity), or implicitly, by fetishizing sweeping rubrics such as dignity, equity, social justice, sacredness, privacy, and consent at the expense of the health and lives of actual people.

The irony is that this line of argument invokes the *idea* of actual lives, while denying what many actual people say: that it's more complicated, that disability only matters in context, and that injustice abounds. But even on his own terms, Pinker's assertions ring hollow. Take the term "social justice," which Pinker derides as "nebulous." Even if we accept Pinker's clear preference for quantitative measures over qualitative ones, one can examine rates of, say, sexual abuse of women with intellectual disabilities, or the differential rates of execution for white and black men, or the pay gap between CEOs and workers in the United States. We can, and always will, argue about what "justice" means, but that does not mean it is unreal or an unworthy goal. In that, it is like other human experiences which, like nebulae themselves, are as real as their borders are uncertain. "Dignity," for example, may be difficult to define, but for many people with disabilities, the assumptions that their lives are less valuable, or defined by suffering, are assaults on dignity as palpable as any physical fact.

As long as humans are talking about anything, we will be talking about how the conversation should go. But Pinker's approach is unusual: his addition to the conversation is an attempt to shut it down. Because his op-ed is linked to a specific biotech application (CRISPR-Cas9), it offers a way into a set of critical questions: How is the conversation about biotechnology itself narrated to us? How are the participants in the conversation described? To what extent does a narrative about that conversation constitute a persuasive pitch for the biotech application in question? And in what ways do ideas about disability underwrite the conversation? Thinking about these questions might help us imagine a better conversation about biotechnology, a necessary prelude to imagining biotechnology's place in the world.

* * *

Though Pinker mentions CRISPR-Cas9 early in the piece, he then shifts to discussing "biomedical research" in general. Conflating the two is misleading: most biomedical research is not controversial. No credible bioethicist opposes a cure for Alzheimer's or pancreatic cancer. What *is* controversial, however, is the use of CRISPR-Cas9 to edit the human germline. It is this, not "biomedical research" in general, that is the subject of proposed moratoria and bans. But Pinker omits this central distinction. That omission is telling: even when we are discussing other biotech applications, from prenatal testing to de-extinction, germline editing is often the shadow subject. But because directing our evolution is categorically different from altering animals, screening future children, or treating Alzheimer's disease, it's critical to see when these are conflated, to see when health is used to leverage improvement. To focus, as best we can, on the blur. That's especially true when the biotechnology itself constitutes a gray area.

At this writing, over forty countries ban germline modification in people. For varied reasons, including opposition to human experimentation, a concern for social justice, and religious conviction, many see genetic engineering, like human cloning, as off limits for any reason. It's a bright line not to cross. But the procedure at the center of this chapter, is—depending on your point of view—either on or over the line. Called *mitochondrial transfer*, *mitochondrial donation*, or *mitochondrial replacement therapy* by its advocates, the procedure combines the DNA of two parents with the mitochondrial DNA (mtDNA) of a third, with the goal of producing a related child free of mitochondrial disease.

When we talk about human DNA, we're nearly always talking about *nuclear* DNA—the (usually) forty-six chromosomes we inherit from our parents, resident in the nucleus of the fertilized egg: the human genome. But we each have a second complement of DNA, inherited only down the maternal line, and resident in the mitochondria: tiny, energy-producing organelles in the cytoplasm (outside the nucleus), often described as the cell's "batteries." Mitochondrial DNA is drastically more limited in reach than nuclear DNA, but mutations can nonetheless result in severe disease.

Mitochondrial replacement therapy is an attempt to avoid that disease. Because some women have disease-causing mutations in their mtDNA, the idea is to find a donor—a second woman, with healthy mtDNA—and construct a fertilized egg, combining the nuclear DNA of the parents and the mtDNA of the donor. This is performed via microsurgery: in a common explanatory metaphor, the "yolk" or nucleus of the donor egg is removed, and the fertilized "yolk" of nuclear DNA is inserted. The hoped-for result is a healthy, genetically related child, albeit one that required three people to produce.

For this reason, some call the procedure *three-parent in vitro fertilization* (IVF). But you can be a parent without contributing genes, and the egg donor in this scenario is unlikely to act as a parent in any meaningful way. Similarly, "mitochondrial replacement" or "mitochondrial transfer" captures the intention of the procedure, but not its complexity. (*Nuclear genome transfer* is technically more accurate, but no one uses it; I prefer *three-person IVF*.)

As with noninvasive prenatal testing and de-extinction, persuasion begins with the name. Just as "noninvasive" tees up a contrast with amnio, and "de-extinction" implies that extinction can be undone, "mitochondrial donation" and "therapy" imply acts of altruism and healing, respectively. "Mitochondrial transfer" suggests that the mitochondria are moved; in fact, the nuclear DNA is moved. "Mitochondrial replacement" sets up a common metaphorical narrative in which "batteries" are "replaced." If the procedure is simply a matter of changing the batteries, which have nothing to do with the "real" DNA we inherit, the procedure's downsides are minimized. But the battery metaphor is both true and misleading. Mitochondria, like batteries, are in fact responsible for energy production in the cell. Unlike batteries, however, they are self-replicating, and unlike batteries, which have two points of connection to the systems they power—positive and negative—mitochondria have complex regulatory functions within the cell. As in other cases, the ordinary descriptive language advances an idealized version of the biotech application in question.

With respect to public opinion, what the technology *is* matters less than how it is *described*. Nonspecialists in a democracy—that is, most of us—have little else to go on. If we believe that NIPT is certain and without risk, we won't see its downsides; if we believe that actual mammoths can really be de-extincted, we're more likely to want to see them; if we believe that, through a simple act of microsurgery, bad mitochondria can be replaced with good and a healthy child will result, we're likely to approve. Popular accounts of technology are the basis of our common understanding. Paying close attention to their rhetoric can help us understand the assumptions in play and the persuasive appeals being made to us. This is particularly true when the account is framed not as persuasion, but as a report.

*　*　*

On June 27, 2014, the *New York Times Magazine* ran a feature article by Kim Tingley, entitled "The Brave New World of Three-Parent I.V.F." Despite its ominous title, the article leans strongly in favor of the procedure: the invocation of dystopia, of a brave new world of reproductive manipulation, is only a prelude to suggesting that fears of dystopia are overblown. The article focuses as much on the controversy as on the procedure, advancing a standard narrative about the conversation: rational, pragmatic scientists are opposed to uninformed, emotional opponents.

The article is anchored by a profile of Dieter Egli, a pioneer and advocate of the procedure; we see him reflecting thoughtfully, practicing science competently, and listening carefully to opponents. Also profiled are families who benefited from cytoplasmic transfer, a crude precursor to the current technique; their heartfelt gratitude is recorded, and the health of their children is emphasized. One of the families appears in a color photograph; there are two photos of Egli in the lab, and another photo showing a magnified cell nucleus pierced by a glass needle, its genome about to be removed. The photos offer a pattern familiar from websites advertising NIPT, a visual rhetoric of science, precision, family, and genes.

No one questioning the technique is given remotely equal narrative weight. Like people with testable conditions in NIPT ads, they exist without pictures, and mostly in the negative. Opposing views are typically represented by fearful anonymous comments; when people are quoted on the record, they are represented in brief snippets. Often, their objections are subsequently answered, giving those in favor of the procedure the last word. For example:

> In the months leading up to the meeting, the F.D.A. received several hundred emails from members of the public objecting to the idea of three-parent embryos on grounds that included: "It's bizarre"; "You are walking in Hitler's footsteps if you allow this"; and "We will have a world of mad scientists."
>
> As the scientists who were pressing for mitochondrial replacement kept pointing out, these fears were somewhat unfounded.

The centerpiece of the article is a public meeting on three-person IVF held by the FDA. Discussing the public-comment e-mails received before the meeting, Tingley's condescension is evident:

> In the United States, it seemed as if the most vocal members of the public felt disturbed by the technique without necessarily being able to articulate why: Of the nearly 250 emails the F.D.A. received before its February meeting, most objecting to "three-parent babies," more than half of them were form letters.

The opposition between reasonable scientists and fearful publics is ubiquitous in popular writing about biotechnology. In "Unnatural Reactions," published in *The Lancet*, Philip Ball dismisses opponents of "mitochondrial transfer," implying that their objections are driven by fear of "the unnatural" or "the unknown." Though Ball notes these objections, he neither engages with their substance nor offers evidence that they spring from fear. Ball also cites the philosopher Russell Blackford, who argues that "appeals against violating nature" come from those who are "searching for ways to rationalize a psychological aversion." At *Slate*, in an article favoring three-person IVF, Jessica Grose quotes a doctor approvingly. "'Every time we get a little closer to genetic tinkering to promote health—that's exciting and scary,' Dr. Alan Copperman, director of the division of reproductive endocrinology and infertility at Mount Sinai Medical Center in New York, told *The New York Times*. 'People are afraid it will turn into a dystopian brave new world.'" In *The Guardian*, columnist Polly Toynbee weighed in on the debate over three-person IVF in the UK: "Stuffed with the religious and rabble-rousers who stir up fears of Frankenstein babies, many in both houses will make noisy speeches, ignoring the science." Taken together, accounts like these sketch the vague outlines of a reliable straw man, a caricature with which to argue. The standard adjectives are *fearful, uninformed, paranoid, Luddite, vociferous, loud, anti-science, anti-technology*; the standard verbs are *worries, frets, complains, fears*. Sometimes the straw man is religious, sometimes politically conservative, and sometimes radical. Ironically, these are literary tactics being used to heighten the authority of science: a character is constructed in order to be dismissed.

That citizens are often uninformed about science, that their reactions are often extreme, is beyond question. But dismissing their reactions does not serve the larger conversation that we need to have, and arguments for new technologies that begin by framing the public as ignorant and emotional are counterproductive. "By constructing the public as ignorant," writes science studies scholar Brian Wynne, "when that public may in its own idiom be expressing legitimate concern or dissent, scientific institutions inadvertently encourage yet more ambivalence or alienation."

It's true that many find three-person IVF repugnant. It's also true that many are "disturbed...without necessarily being able to articulate why." It'd be more useful to ask *why* people feel this way than to dismiss them out of hand: to ask what values drive the repugnance, and to reject the self-serving frame that the scientists are the only reasonable ones. In their 2015 article "CRISPR Democracy: Gene Editing and the Need for Inclusive Deliberation," Sheila Jasanoff, J. Benjamin Hurlbut, and Krishanu Saha write,

When larger questions arise, as they often do, dissent is dismissed as evidence that publics just do not get the science. But studies of technical controversies have repeatedly shown that public opposition reflects not technical misunderstanding but different ideas from those of experts about how to live well with emerging technologies. The impulse to dismiss public views as simply ill-informed is not only itself ill-informed, but is problematic because it deprives society of the freedom to decide what forms of progress are culturally and morally acceptable.

If the only questions are technical, then the scientists' voices should have priority. But as Hurlbut writes elsewhere, the strict focus on technical issues is insufficient. Beyond the details of any application are questions beyond the technical, "longstanding questions about what features of human life ought not be taken as objects of manipulation and control. They are questions about our responsibilities to our children and our children's children, where the mark of our actions will be inscribed upon their bodies and their lives."

Hurlbut traces the narrow technical focus of our current conversation to Asilomar, where, in 1975, a group of scientists met to draw up restrictions on recombinant DNA—the then-new ability to import genes, using viruses, from one organism to another. Reading against the standard story of "transparency and openness," Hurlbut notes that the conversation both empowered the scientists and disempowered the public. By framing the conversation in narrow terms, on "technical questions about risk assessment," Asilomar's participants excluded key questions: "[how] these technologies might be used and what those uses might mean for our lives"; by restricting the conversation to experts, the public's role was "passively learning—and deferring to—science's authoritative judgment about what is at stake and when a democratic reaction is warranted."

In "The Brave New World of Three-Parent IVF," Asilomar is invoked for a very different reason. The paragraph about Asilomar is preceded by a rhetorical question—"So what, exactly, are we so afraid of?"—and begins with an answer: "We know the double helix as our identity; we take personally the thought of tampering with it." Tingley then quotes a bioethicist who argues that scientists at Asilomar, in their excitement for recombinant DNA, made the technology seem intimidating: "[T]he public was titillated but also frightened." The paragraph concludes with a clear implication that fear was unwarranted: "In the end, the researchers did splice DNA; this has [led] to many benefits, but not, as feared, a cancer-causing superbug."

It is better to have some discussion rather than none. But the critical distinction is between genuinely useful discussion and its saccharine version, and so far we have had the second more than the first. I do not think that we will have the first, because it means that the people who stand to profit from the technology will have to surrender power: to acknowledge that the *main* stakeholders in technology that affects the species are members of that species, and that insofar as the public is imagined in token terms and judged in terms of their compliance, the conversation may ratify injustice and not address it. If the public exists solely to "be informed" and "be heard," the conversation will accomplish little, and it may even do harm by displacing more meaningful exchanges. Or the conversation may be an exercise, a substitute for action. Having a voice is different from having a say.

* * *

There are some issues—the effectiveness of vaccines, for instance, or the causes of climate change—on which scientific consensus is virtually unanimous. This is not the case for three-person IVF, where scientists strongly disagree. For this reason, the frame of Rational Scientists versus Fearful Rabble does violence to both sides, because it oversimplifies the questions raised *by* scientists, implying unanimity where none exists. The case against moving three-person IVF to the clinic was summarized by stem-cell biologist Paul Knoepfler in an open letter to the parliamentary committee in the UK then considering approval of the procedure. Knoepfler did not oppose the procedure in principle but argued that more study was needed, and that the procedure might *cause* illness or death, rather than preventing it. Citing work published in *Science, New Scientist,* and elsewhere, Knoepfler argued that harm might occur in a number of ways, including the combination of mismatched nuclear and mtDNA; "the impact of mitochondria on traits" (contrary to Tingley's account, mtDNA appears to regulate nuclear DNA); and "the preferential replication of even tiny amounts of carryover mutated mtDNA." Knoepfler continued, "What are the alternatives? Preimplantation genetic diagnosis (PGD) is a powerful technology that can help many (although admittedly not all) families dealing with this situation in an effective manner." (In PGD, fertilized embryos are tested, and only healthy embryos are implanted.)

Knoepfler was careful to point out that he was not a "[L]uddite," and that he was open to the use of the technique in the future. His point wasn't that the technique was scary; it was that our knowledge is incomplete. If mtDNA and nuclear DNA are more intricately linked than we now suspect, then the experiment's descendants, instead of being healthy, could pay the price for what we don't know. To even find out in a truly scientific way would involve experimenting on human beings—and if we *really* want to know, we'd need to have a large sample size, and track the resulting humans through more than one generation. (Effects could potentially appear in later generations, which means that we'd not only be experimenting on the children who didn't consent, but on their children as well.) For the procedure to be successful would require both flawless technical execution and sufficient knowledge of the cell's workings, neither of which can be realistically assumed.

Since 2014, the story of three-person IVF has been one of contrary motion: even as scientific doubts have intensified, the technique has moved closer to implementation. In February of 2015, the UK granted regulatory approval to the procedure. This entailed an act of definition, or redefinition: three-person IVF was declared not to constitute human germline modification, which is banned by law in the UK. In May of 2016, Dieter Egli's lab published an article in *Cell Stem Cell* that confirmed one of Knoepfler's warnings: that the small number of defective mitochondria brought along with the transferred nucleus can, in the newly assembled cell, "outcompete" the resident healthy mitochondria. This result, according to Egli, "would defeat the purpose of doing mitochondrial replacement." In June 2016, a paper in *Nature* from Mary Herbert's lab in Newcastle, UK, announced that pronuclear transplantation (PNT)—the very form of "mitochondrial donation" on which regulatory approval had been based—"[was] not well tolerated by normally fertilized zygotes." Or, as Jessica Cussins noted in a blog post at *Biopolitical Times*, "It did not work."

In the same paper, the researchers described a different form of PNT that worked better, but that still resulted in carryover of defective mitochondria. Incredibly, as Cussins wrote, most media outlets reported the story as an advance, not a setback. (Her proposed alternative headline: "Scientists effectively distract press and public from the dangers inherent to the controversial technique they promised wasn't unsafe by introducing a new technique promised not to be unsafe.") Paul Knoepfler praised both papers but also wrote that they "clearly indicate that the field is not ready to use this technology to create actual people. It would be reckless to do so now without getting more data first." In November of 2016, Shoukhrat Mitalipov, an Oregon scientist who argues for using the technique to treat infertility, published a paper in *Nature* that also showed diseased mitochondria could outcompete healthy ones; as Karen Weintraub reported in *Scientific American*, "[T]he new laboratory study shows that the mother's mitochondria can sometimes replicate faster than the donor's and come to dominate again, potentially bringing disease with it."

In December of 2016, the Human Fertilisation and Embryology Authority gave final approval for the clinical use of the technique in the UK. But by that time, the first baby using three-person IVF had already been born in Mexico, in April, to a Jordanian couple whose two previous children had died of Leigh's disease, a mitochondrial disorder. Five embryos had been created for them in the United States by John Zhang of the New Hope Fertility Center; one did not develop, and three were aneuploid (had extra chromosomes), leaving a single viable embryo. This embryo was transported to Mexico, where the procedure is legal, and implanted in the mother at New Hope's clinic in Guadalajara. Though the child, a boy, is reportedly healthy, his cells showed evidence of mitochondrial carryover; his parents have resisted permission to follow up on his health as he grows. Around the same time, news arrived that two pregnancies based on three-person IVF had begun in the Ukraine. These were not meant to avoid illness; they were experimental procedures, meant purely to address infertility.

At this writing, the child born in Guadalajara is prominently featured in the New Hope Fertility Center's online marketing. On their website, under the headline THE NHFC DIFFERENCE, the first item is "First live birth using human oocytes reconstituted by spindle nuclear transfer for Mitochondrial DNA mutation causing Leigh syndrome." There's a checkmark beside it, like something on a human species to-do list. The site's aesthetic is reminiscent of the world of NIPT—stock pregnant model embraced protectively by stock handsome man, a color language of feminine purple and clinical blue, images of microscopes and test tubes connoting authority, precision, science—but the New Hope Fertility Center website, at this writing, has yet to reach the refined, understated heights of natera.com or sequenom.com. They are different in the way a used car lot and a Lexus dealer are different. The center's website features a stork above its banner headline; the circular payload slung from its beak forms the letter O in Hope.

> ARE YOU LOOKING FOR THE BEST FERTILITY DOCTOR? CALL US AT (555) 555–5555. Proudly serving Patients speaking **Chinese**, **Arabic**, and **Spanish**. FAMILY STARTS WITH HOPE: Click **here** to schedule your consultation. IVF FOR EVERYONE: Now offering payment plans as low as $40 per month. WORLD LEADER AND PIONEER OF MINI-IVF™.

To me, the website's pitches show the limits of sober, reflective public conversation. Almost everyone thinking about the impact of biotechnology agrees: more discussion is needed. But the truth is that what actually happens will be driven by what people can afford to do in a globalized economy. For that reason, considering the risks and benefits of three-person IVF as a medical procedure, or the conceptual difference between healing a child and creating one, is only the beginning. How will market pressures affect the science that supports the procedure, the language with which we speak about it, and the way the procedure is done? Will the profit motive drive the technique beyond unambiguous instances of disease—as is already being done in the treatment of infertility? If the technique is scaled up and becomes common, who will provide the eggs, and how will we avoid either commodifying women's bodies by paying for eggs, or having egg donors bear the risks of retrieval? Given the expense of the technique, who will have access to it? And if, down the line, something goes wrong—if, for example, the child, or the child's child, becomes ill as a result of the procedure—who will be responsible? Stem-cell biologist George Daley told *Nature*, "Going to Mexico is a way to evade the stricter regulatory regime in the US and UK.... The danger is to the families and the infants who are being born with this procedure. They're taking all the risk before really being fully aware of the success rate and failure rate."

So it was troubling to hear leading scientists argue that Zhang's hand had essentially been forced: that because of burdensome regulation, he had simply *had* to go elsewhere. Shoukhrat Mitalipov, in a written statement provided to *Nature*, "said that the strict US regulations are forcing researchers to move trials into other countries." Dieter Egli, speaking to *Nature* in late September 2016, said that the procedure "was not well done," but that October, he said, "I think the scientists and doctors showed a lot of courage exposing themselves to this type of criticism in doing this treatment... I think there is an imbalance in regulation and oversight in some places, putting novel treatments on the long bench, and therefore it had to be done that way." Zhang himself said, "We'd love to do it with partners around the world and reach out to more families that might need help."

* * *

Three-person IVF is a likely precursor to inheritable germline modification in people. For this reason, it seems reasonable to expect that, as human germline modification draws closer, feature articles like "The Brave New World of Three-Parent IVF" will accompany the possibility. Perhaps one will feature a child born in Mexico in 2017, via three-person IVF: if he grows up to be healthy, perhaps he, or a girl born in Ukraine, will serve as implied proof for the new technique, one more example of something the Luddites worried about and got wrong, or perhaps the child will be used to say that we have already crossed the line anyway, so we might as well go ahead. Each advance is used to leverage the next: *there was no cancer-causing superbug; the children born to cytoplasmic transfer turned out okay.*

Discussions of NIPT, de-extinction, Synthia, three-person IVF, CRISPR-Cas9, and the like all have human germline engineering as their subtext, even when they do not invoke it directly. Of course, in published discussions—including this one—the prospect of that future, of a "brave new world," is a selling point. "Was the future of our species being decided at this Hilton," Tingley writes, "with its seaweed-colored carpet and free packets of perfumed soap, or were we simply weighing the evidence for supervised tests of a specific technology with definable limits? Or both?" It may be that all this discussion serves to normalize the possibility, even when it is meant to warn.

Near the article's end, Tingley brings that subtext to the fore, addressing human enhancement directly:

> What often gets lost in the loaded language of the debate over three-parent babies is the fact that ordinary human reproduction is, by definition, genetic modification. The risks involved are unpredictable and potentially tragic; the subject of the experiment is a future person who cannot consent. We constantly try to control this process, to "design" our children, starting with our choice of sexual partner. During pregnancy, we try to "enhance" them by taking folic acid, not smoking, avoiding stress; once they're born, we continue the process with vaccines and nutritious food, education, clean air and drinking water. Some of these pre- and postnatal environmental factors, we now know, change their biology in heritable ways. Is mitochondrial replacement, because it takes place in a petri

dish, any more unnatural or morally repugnant than this? Would the answer change if the technique turns out to cure age-related infertility in addition to preventing disease?

All of these points are familiar from the debates over human genetic engineering, and they are all from one side of the debate. We're *already* "designing," *already* affecting human generations, so why not be more precise and deliberate? It's almost a shock when the paragraph's last two sentences mention "mitochondrial replacement," since we've clearly been talking about enhancement all along. The discussion of one slips easily into the discussion of another. As the paragraph moves forward, its subject mutates: *reproduction, modification, experiment, "design," "enhance."* These swift changes are enabled by rhetorical tactics we've already seen, including the Analogy to the Familiar: something new (genetic modification) is defanged by comparison to something old (Ordinary Human Reproduction). The advantages of this tactic are obvious. If the new thing and the old thing are effectively identical, then the new thing shouldn't be scary: it's already here. But the analogy between Ordinary Human Reproduction and genetic modification is tenuous. In Ordinary Human Reproduction, genes are combined from two people, not three, and the process does not require microsurgery, embryo biopsies, and an offsite fertility clinic in Mexico beyond the reach of the FDA. "Modification" is a red herring; at issue is "engineering."

It's also questionable to equate "choosing a sexual partner" with "design." Tingley appears to assume a *heterosexual* partner, as is typically the case in NIPT marketing, but choosing a sexual partner may have nothing to do with reproduction, let alone design. And to equate procreation with design ignores other ideas about parenting. One can, for example, believe that parenting is a matter of accepting, more than designing, and that trust in each other's values, and in the ability to work things out, will always trump "design." In fact, the child may be welcomed precisely because she is *not* designed, because she is her own person, a gift. One may also feel that the genetic qualities of each individual are of less importance than the qualities of the connections between individuals—in a family, in a community—and that if those fail, the best-designed attributes cannot succeed.

Neither can pre- and postnatal care be equated with "enhancement." In discussions of human modification, "enhancement" typically means *adding* something not otherwise possible: better memory, bigger muscles, improved memory, and so on. "Enhancement" is usually contrasted with therapy, so to compare enhancement with smoking cessation, stress avoidance, or good nutrition is a stretch at best. These are ways of nurturing an existing process, not improving the developing embryo. Taking folic acid is closer to the mark, but this, too, is meant to avoid impairing an existing process of development—which is categorically different from direct alteration of the DNA or mtDNA, a process which *precedes* development.

It is true, as Tingley writes, that "[s]ome of these pre- and postnatal environmental factors...change...biology in heritable ways." The implication favors three-person IVF, and even genetic modification: we're already doing it informally, so why be afraid? But an alternate answer is to say that if this is so—if good prenatal care, clean water, and better nutrition *do* influence future prospects—then we should reconsider the balance of high tech and low. We should consider our relative investment in cheap, side-effect-free, tested, equitable measures as opposed to experimental, expensive, high-tech ones. It isn't an either–or: we need high tech and low. But we are entranced by the new, perhaps by risk itself, and that fact drives not only which technologies we pursue, but the conversation about them. The economy of attention in which we live favors what's novel and exciting: the possibility of altering the species' future is sexier than the possibility of clean water, clean air, and good nutrition, and cutting-edge technologies are inherently exciting to many. But the technology itself matters less than the principles it embodies, the goals it is meant to serve, and the assumptions by which it is justified.

In writing aimed at the lay public, those assumptions often take narrative form. Discussions of biotechnology often take the form of fragmented morality tales: arguments with characters mixed in, appropriate for a time when life is insistently fragmented and reassembled. Medieval morality plays featured allegorical characters: Everyman, God, Death, Good Works. The new tales also offer total pictures of the world, but their characters include the Scientist, the Luddite, and the Designer Baby. They also include Disease, Disability, and Death, but in some of the new stories, these are less givens of life than terms subject to revision, with disease conquerable and with death forestalled or vanquished altogether. These outcomes, these promises, are only possible because Everyman is himself subject to revision.

In the article's conclusion, we see Dieter Egli extracting a nucleus from a cell. He's reframed as an icon of scientific discovery, an astronomer:

"[H]e flicked on the monitor and the egg appeared, a luminous round moon, as if his instrument were a telescope aimed not at a single cell but at the night sky." Egli extracts the egg, and the genome is described as all-powerful: "[T]here in the needle, was the genome, ready for transplant: an oblong seed, arbiter of all life." Earlier in the article, that power is linked to the nuclear family, and to intelligence. Profiling the family that used cytoplasmic transfer (the crude forerunner of mitochondrial trans-fer) in 1996, Tingley writes, "As far as anyone knows, mitochondrial DNA (mtDNA) governs only basic cellular functions; [Maureen] Ott understood that her and her husband's nuclear DNA would *determine* their child's characteristics—height, eye color, intelligence and so on." (Italics mine. Height can be objectively defined, and intelligence cannot; both, however conceived, are the result of complex interactions between genes, epigenetic regulation, and environment.)

But as written, the narrative embodies the link between genes and intelligence—or a performance of the specific sort of intelligence valued in this culture, at this time. Maureen Ott's daughter Emma, we're told, "gets straight A's, is senior-class treasurer and plays varsity sports." Another girl born after the same procedure is described as "exceptionally bright and healthy," and her mother is quoted as saying, "As my daughter grew and she's fine and so intelligent, it just backed my belief that it [cytoplasmic transfer] was the right thing to do." But then, the performance of intelligence is celebrated by the concluding image of the article: a scientist, contemplating the celestial object of the egg, extracting from it "the arbiter of all life."

Granted that awesome power, the genome is only the brightest star in a familiar constellation: genetic determinism, the importance of intellect, the nuclear family, and the power of science to improve us. An ancient pattern, its light still shining on us, from long ago.

9 Model Worlds

We used to kill time, playing *Life*. When Laura was nine, it was her favorite board game, probably for its number of tiny parts and the amount of time an average game consumed. If we were playing *Life*, we were, by definition, not occupied with the ephemera of life; we were fully present, or mostly so. I'd fit the first-trimester-sized pink and blue pegs into a plastic car, spin the plastic wheel, and trundle through a generic story of weddings and mortgages and tuition. I'd glance toward the laptop and then glance away.

Every board game constrains the world. In *Clue*, the players roll dice like gods above a crime, tracing the threads of fate with golf pencils. Laura liked *Clue*. *Chutes and Ladders* was easier to understand and better for working on counting, but the game is boring, merciless, and unjust. As soon as you clamber ahead of everybody, you slide back down; at the end, you can wait for turn after turn before rolling the exact right number to win. Also, you have to switch directions when counting—left to right and then right to left—when one direction is hard enough. Which is how we came to *Monopoly Junior*.

If *Clue* abstracts murder, and *Life* abstracts life, and *Monopoly* abstracts capitalism, then *Monopoly Junior* abstracts the abstraction. It is a model of an artificial world. In his great essay "The Search for Marvin Gardens," John McPhee alternates scenes from a fierce, rapid, mano-a-mano game of Monopoly with scenes from the real Atlantic City: we understand that the impulse to compete and dominate gave rise to the actual urban blight transformed, in the game, into plastic hotels, property cards, and silver hats. In *Monopoly Junior*, Atlantic City has been replaced with an amusement park. Purple is the Balloon Stand. Blue is the Echo Canyon Coaster. Jail is Lunch. Free Parking is Free Time. Houses and hotels are replaced by Ticket Booths. Luxury taxes, assessments per house, mortgages, all gone. It's two dollars to see the Water Show or the Fireworks. The lowest denomination is one dollar, the highest five. There is a single die. There are no utilities, and the railroads have been scrubbed of place, toil, and capital: they are simply Red, Blue, Yellow, and Green, and they cannot be owned. If you land on one, you just roll again. We liked *Monopoly Junior*. It was a manageable, Laura-sized world. We played until the board filled up with Ticket Booths, or Laura won, whichever came first. I rarely won: Laura was the Dale Carnegie of unreality. She'd land on Free Time and gather the bad luck of others, the taxes on others' misfortunes. She'd accumulate Ticket Booths, draw lucky cards from Chance.

The games were models of life: they granted endless possibility within severe and artificial limits. But I did not know if the models' assumptions would hold for Laura, or what possibilities she might have. What would *her* life be like? What relevance did they have, these abstracted nuclear families, taking their winding journeys in plastic cars, acquiring pink and blue children; how would Laura fare in the amusement park of capitalism, its random movement from one entertainment to the next, its accumulation of symbolic wealth? What happens when the child outgrows the game? What we told ourselves then was true: Laura was acquiring skills and language, learning to interact, but as we approach the cliff of eighteen, as Laura's school career comes to an end, I know that the preparations we make are necessary and insufficient. Laura loves school more than anything: her high school is a model in every sense, both a distillation of society and a shelter from its dangers. But the model is not the world.

*   *   *

The scientists are playing life. In mouse models, in cell lines, in sequencers, beakers, petri dishes, in algorithms that create virtual children or predict diagnoses for future ones, in synthetic cells that quote Joyce, the molecule that becomes us is sequenced, altered, copied, studied, compared. In vitro, in silico, in vivo.

In 2013, researchers in Jeanne Lawrence's lab, at the University of Massachusetts, reported that they had found a way to "silence" the extra chromosome associated with Down syndrome in vitro—that is, to prevent the genes of the extra chromosome from being expressed into proteins. Because people with Down syndrome have three copies of each gene instead of two, their proteins are overexpressed, which alters both development and function. Lawrence had adapted the mechanism that prevents overexpression in female mammals: since females inherit two X chromosomes, one is automatically turned off by a gene known as XIST. The gene codes not for a protein, but for a long stretch of RNA. That RNA, as science writer Ed Yong explains, "coats one of the two X chromosomes and condenses it into a dense, inaccessible bundle. It's like crunching up a book's pages to make them unreadable and useless." The gene's name is pronounced "exist."

To silence the extra chromosome, the researchers inserted copies of XIST into cells taken from a boy with Down syndrome. The boy's cells had been "reprogrammed" to behave like pluripotent stem cells—that is, cells that can potentially become any kind of cell, as opposed to cells whose developmental fate is already decided. The cells were then cultured, and the XIST gene was inserted with zinc finger nucleases, a genome-editing tool that predates CRISPR. The experiment worked; the book of the extra chromosome was rendered unreadable, and as the cells divided, they developed more normally. The boy was present in the model, as code, but some of his code had been silenced, and his cells had become something else.

Chromosome silencing is a suggestive term, particularly for a population so often silenced and unheard. It's tempting to run with the metaphor, to say that silencing occurs at the molecular level and the societal—which, in many ways, it does. But in this case, it's more complicated. As basic research on epigenetic mechanisms, the XIST project has applications beyond Down syndrome; unlike a prenatal test or an algorithm that generates virtual children, it's not focused on selection. By definition, the project implies a treatment-oriented approach: if the condition is purely something to eliminate, then the only research needed is in optimizing prenatal detection. Lawrence hopes for clinical applications one day, though that day is far off. She also hopes to better understand the complicated biology of Down syndrome, the pathways between a trisomy and a person with a syndrome. (Lawrence also regularly invites parents of children with Down syndrome to speak to her medical school classes.)

And yet the account of the discovery's significance reveals a familiar set of contradictions. In the conclusion, the authors look ahead to possible treatments: "Our hope is that for individuals and families living with Down's syndrome, the proof-of-principle demonstrated here initiates multiple new avenues of translational relevance. …" But the abstract opens in another key altogether: "Down's syndrome is a common disorder with enormous medical and social costs, caused by trisomy for chromosome 21."

In one frame, people with Down syndrome are patients: they have a disorder that can potentially be treated. In another, they are a burden: the disorder has "enormous medical and social costs." The two sit uneasily together, an old and familiar tension. The common thread between them is the medical model: Down syndrome is understood as a disease, not variation, so other aspects of the condition are silenced. As the boy is reduced to (reprogrammed) cells, so the condition is reduced to the ideas of *cost* and *disease*.

This is partly just the boilerplate of scientific persuasion. In the increasing competition for a shrinking pool of grant money, scientists have to explain their work in economic terms. In a grant application to study ovarian cancer, for example, it's a standard move to note cancer's societal costs. If this is the case, then the money spent on the research is a worthwhile investment. But the same argument has a different resonance, a different history, when attached to human groups. To see cancer as costly is different from citing Down syndrome as costly, because Down syndrome tends to be equated with the people who have it, and because we have a long history of citing the expense of intellectual disability as part of an argument for elimination.

In the 1970s and 1980s, writes Alexandra Minna Stern, prenatal testing was justified on eugenic and economic grounds: it would "not only eliminate supposedly deleterious genes in future generations but save millions of dollars by reducing rates of institutionalization and producing more industrious citizens with higher overall earning power." In fact, as Robert Resta has shown, the iconic number 35—the age after which pregnant women were advised to pursue amniocentesis—was driven not by a neutral risk assessment of miscarriage, but by economic concerns. The studies in question, writes Resta, "asked the same question—At what maternal age is the economic cost of amniocentesis lower than the economic cost of caring for people with Down syndrome?" And yet the studies underlying the age recommendation were themselves flawed:

> All of the studies compared the cost of amniocentesis with the current medical, custodial and educational costs of caring for individuals with Down syndrome. No study compared the cost of amniocentesis to the cost of developing better programs to improve the medical care, economic opportunities, and education of people with Down syndrome.

Given this long and troubled history, the justification of chromosome silencing with "enormous medical and social costs" is deeply problematic.

Doing so aggregates a population into a negative; it translates a micro-scopic measure into a macroeconomic one; and it traces all the costs to the group, and not to the society that has failed to welcome it. Between the chromosome and the cost, the molecular model and the economic, there's no room for actual lives, or for the idea that Down syndrome is a form of human variation, an acceptable way of being in the world.

*　*　*

Disability is widely feared, and widely misunderstood, and these fears have historically taken economic form. Early-twentieth-century eugenicists lamented the cost to society caused by the "feeble-minded" and hoped to save the money wasted on them for the more deserving, not only in the present but also for future generations. It is easy to see, with the benefit of distance, the prejudice and disdain informing their calculations. It is easy to see that cost was a proxy for ideas about human value, that ideas of value are embedded in culture, and that our culture is perennially troubled when it comes to thinking about disability. But similar notions are only a com-ment section away:

> I'm responsible for the costs and consequences of my lifestyle choices, and those who choose to procreate should be held accountable, as well. To deliberately and knowingly bear offspring that will present abnormally high costs to society is not a virtue or act of selflessness, it's extremely selfish and short-sighted. And as usual, the village will be expected to pick up the tab.

And another:

> Would-be parents wouldn't be so eager to roll the dice if they knew they would be responsible for ALL the costs of the extra medical care and therapy that dis-advantaged children would need, as well as the creation and financing of a trust fund to take care of these individuals once the parents have died. But it's easy to be altruistic on somebody else's dime, isn't it?

These were sent to me, in 2013, by Alison Piepmeier—the friend who could not join us at the NSGC meeting in New Orleans because she was undergoing chemo. The comments followed a piece Alison wrote for the *New York Times* entitled "Outlawing Abortion Won't Help Children with Down Syndrome." Alison had turned the comments into a project, study-ing them, looking for patterns. "I just treat it as research," she told me. In fact, I think they disturbed her deeply, but she could only look away by looking closer, by working as a scholar and trying to salvage something useful.

You cannot predict what any one individual, over the course of a life-time, will cost. You can have chromosomally typical, intellectually gifted men who craft Ponzi schemes, devise unsustainable financial instruments, or plunge countries into deficit-exploding wars. (Most of our large-bore societal costs are brought on by those fully able in mind and body. There is no gene for opportunistic assholes, and even if there were, I would not support a population screening program around it, but if we're looking to save money, perhaps we should look elsewhere than to people with disabilities.) At the same time, the high unemployment rates among people with disabilities, and the legal payment of subminimum wage, are due in part to a failure of imagination and to a story dominated by deficit and stigma.

*   *   *

In his seminal history *Inventing the Feeble Mind*, James Trent writes that in the late nineteenth and early twentieth centuries, those institutions that had not yet given up on their educational mission were fond of an instructional method called the "object lesson." It was a model of the world outside, a game of life: in it, "information about community life was conveyed through the use of miniatures." I have looked for pictures but cannot find them. I imagine sad *Antiques Roadshow* versions of the toys Laura played with as a child: little houses, little people, little stores. As Trent notes, residents were learning about a community they were unlikely to rejoin. He reproduces a letter written by a girl to her parents: "My dear mamma and papa and aunt Carrie, I would not mind comming back home at once." The letter was never delivered. It was common to discourage parents from contacting their children after they had been placed; there were also parents who did not wish to be contacted, who found home visits disruptive, who wanted to move on.

The object lessons were models inside a model: the institutions were miniature versions of the society they served, warped versions of the world beyond, with their own castes, laws, procedures, and limited freedoms. They existed for many reasons: parents, unable to handle their children or get help in the community, wanted them; ambitious superintendents wanted to expand them; they answered to a eugenic belief that the "feeble-minded" were a menace, that the "moron" might pass in society, reproduce, and fuel both social cost and crime. They answered disability's question with separation; in so doing, they kept public space clear of the feeble-minded, keeping them out of sight. Invisibility was their chief product: they turned stigma into absence. That motive was made explicit in the founding of Oregon's Fairview Training Center: in a history of that institution, Philip Ferguson cites a report that "strongly recommended that the State try to locate the new institution in 'a secluded valley, upon whose sunny slopes these simple people might dwell away from the public gaze.'" The main goal was separation: the "simple people" were to be *secluded*, *away*. There is a faint trace of goodwill toward the "simple people," but this seems notional at best, like the idea that a Western Oregon valley might have reliably sunny slopes.

Fairview, less than an hour north of my house, was investigated in the early 1980s and finally shut down in 2000. Though the investigation found that inmates had been beaten and neglected, though the guards chained some inmates to seventy-pound blocks and made them drag the blocks up and down the halls as punishment, it is possible that in the institution's early days, the lives of its residents were not completely miserable. They seem to have enjoyed crafts: as Ferguson writes, "From the beginning, the students' handiwork (basketry, darning, weaving, sewing, making pillow lace) was put on annual display at the State Fair, where it 'received much attention … and much to our gratification we carried away a number of blue ribbons.'" Whether the residents themselves got to visit the fair, or whether they were represented only by their handiwork, is not clear. What is clear is that people with disabilities have often been subject to a fiercely conditional welcome. Perhaps Oregon fairgoers cooed over the baskets and lace. Perhaps they felt a vague sense of relief that the people they did not wish to see in daily life were nonetheless provided for and enjoying themselves, that the domestic arts were displayed, that the female inmates could be ladylike without reproducing. But these artifacts of domesticity came from a place that was, in the end, the opposite of a home.

To imagine what it was like to raise a disabled child in 1927 is, for me at least, harder than imagining a world of human upgrades. What parents wanted varied from case to case, and their wants were constrained by imagination and the world, by available choices and thinkable thoughts. Some wanted institutional placements they could not get. Some left children behind and never looked back. Some were torn and changed their minds. One Oregon family, writes Ferguson, had been pressured by the Board of Welfare, on grounds of cost, to place their twelve-year-old twins at Fairview. Their decision to retrieve them occasioned a typical sequence of anguished requests and imperturbable replies. The father writes, "We do not [want?]

them to stay thaire eny (sic) longer than we can take them," and the superintendent replies as follows:

> Dear Sir,
>
> Since you were here Saturday I have discussed with members of the Board
> [of Welfare] your taking Blanche and Hazel out of the State. It was decided that
> if you will take them out of Oregon and keep them out we will be willing to let
> them go. When you come for them it will be necessary for you to sign a permit,
> and it is understood that should you come back the girls will have to be sterilized.
>
> Yours very truly,

*   *   *

A story is a model of the world, a game of life. It distills life in the dish of
language, turning it into sequence. But a story about Down syndrome can
be a form of invisibility: when the condition is narrated in terms of genes,
or abnormality, or cost, the people tend to disappear. This is a particular
problem in some books about genes and disability, in which the voices of
people with other conditions are included, but the voices of people with
Down syndrome are omitted. For some influential cultural voices, intellectual disability appears to be *uniquely* unimaginable, uniquely unreal, even
when the portrait is nominally sympathetic.

In *The Gene: An Intimate History*, Siddhartha Mukherjee's account of
Down syndrome fits this pattern. Mukherjee is an elegant and thoughtful
writer, a deft explainer of genetics. But his account of Down syndrome
is inconsistent with his book's focus. Elsewhere in the book, he emphasizes
variability: genes are not destiny, but part of a system. And in his stories
about mental illness in his own family, Mukherjee shows us that genes are
not only mechanisms; their effects are processed as experience, converted
into narrative, given meaning in individual lives. People with Down syndrome, however, are discussed as a group, and Mukherjee ascribes their
characteristics to genes. Even though he correctly notes that the syndrome
is variable, he presents that variation as purely genetic, as if social factors
were irrelevant.

It's an odd argument for a physician to make. In the last thirty years,
the life expectancy for white people with Down syndrome in the United
States has gone from twenty-five to sixty, due largely to the advent of heart
surgery and antibiotics. Like the gulf in life expectancy between black and
white babies with Down syndrome, it is as pure an instance of environment as you can find. An extra chromosome is an extra chromosome, but

the people who were once considered a burden, then a menace, then fit only for institutions, are now included in general education classes, though hardly often enough. This is culture, not genes. Mukherjee also advances Down syndrome as an instance of genetic influence on behavior, retailing the old stereotype that people with Down syndrome are universally sweet.

The same pattern is evident in Andrew Solomon's *Far from the Tree: Parents, Children, and the Search for Identity*. In it, Solomon writes that "loving our children is a test for the imagination." Over the seven hundred or so pages that follow, Solomon profiles the sorts of imaginative tests that most parents never face. The chapter titles include Deaf, Dwarfs, Autism, Crime, and Transgender. He speaks to parents of musical prodigies, Rwandan mothers who keep the children of their rapists, parents of children with multiple severe disabilities. But as he discovers, "most of the families described here have ended up grateful for experiences they would have done anything to avoid." Solomon is a superb writer: elegant without fussiness, lyrical without pretense. The families' stories are interwoven with extensive research and bookended by Solomon's own quest for identity, first as a gay man seeking his parents' acceptance, and then as a father. Solomon is eloquent on the intersection of different marginalized identities: he writes, "I thought that if gayness, an identity, could grow out of homosexuality, an illness, and Deafness, an identity, could grow out of deafness, an illness, and if dwarfism as an identity could emerge from an apparent disability, then there must be many other categories in this awkward interstitial territory. It was a radicalizing insight. ... Difference unites us."

And yet for Solomon, Down syndrome remains more illness than identity: the people with the condition are never heard from directly. (This is also true for the chapter on autism.) The absence is conspicuous, and it's filled by group characterizations: "People with Down syndrome are often warm and sociable, eager to please, and free of cynicism. Larger studies indicate that many people with Down syndrome are also stubborn, defiant, aggressive, and sometimes disturbed." These sentences also make perfect sense if you subtract the phrase "with Down syndrome," but in any event the two studies Solomon cites are equivocal: they emphasize how little is still known, how people with Down syndrome fare well compared to other children with intellectual disabilities. One study notes relatively *low* rates of true "aggression," like fighting. Higher rates are only present if "disobedient," "argumentative," and "demanding attention" count.

In a widely shared TED Talk, the novelist Chimamanda Ngozi Adichie warned of "the danger of a single story": "The single story creates stereotypes, and the problem with stereotypes is not that they are untrue, but that they are incomplete. They make one story become the only story." She continues, "The consequence of the single story is this: It robs people of dignity. It makes our recognition of our equal humanity difficult. It emphasizes how we are different rather than how we are similar." In Adichie's formulation, the single story is imposed from without, and though she was speaking about Western perceptions of Africa—the way the diverse people of a vast continent are reduced to a "single story"—her message is relevant to intellectual disability as well. The single story emphasizes group characteristics, and in so doing erases the particulars of identity and circumstance; it is a form of invisibility.

*   *   *

In July of 2013, I found myself thinking a lot about the invisible. I had traveled to Denver for the annual National Down Syndrome Congress convention, and even given the fact that Down syndrome lands in the news with regularity, there were two big pieces of news during the week I was there. The first was the report, from Jeanne Lawrence's lab, that the XIST gene had been used to silence Down syndrome in a dish. The second was the release of a full report of the facts surrounding the death that January of Ethan Saylor, the twenty-six year old man with Down syndrome who—on refusing to leave a movie theater for a second showing of *Zero Dark Thirty*—was dragged from the theater by three off-duty sheriff's deputies moonlighting as mall security, handcuffed, and held face down. His larynx was fractured, and he died of asphyxia. Before he died, he was crying in panic for his mother, who was en route to the theater. His death was ruled a homicide by the coroner, but a grand jury declined to indict the deputies, and an internal investigation found none of them at fault. The case received national attention, but the outrage was mainly confined to what we would call, for lack of a better word, the Down syndrome community—those with Down syndrome and those who care about them. It took months for the full record of facts to be released; now, after years in court, Mr. Saylor's family has settled for $1.9 million with the deputies, their employers, and the State of Maryland, none of whom admitted fault.

In Denver, for a few days, people with Down syndrome were unusually visible—not in the context of human interest features or stories about new prenatal diagnostic tools, which is the way they usually appear, but as family members of all ages. Laura was back home, though I had two suitcases full of copies of her story: I was selling my book about her, the one she calls "her book," in the fluorescent, sunless caverns of the Colorado Convention Center. But throughout the center, and to a lesser extent in the streets and hotels nearby, people with Down syndrome were visible everywhere.

There are two kinds of invisibility: a literal one, abetted by new technologies, whose likely effect is to ease a population from sight, and a cultural invisibility, fueled by misunderstanding and fear. The two are interrelated. Misunderstanding and fear can fuel the demand for prenatal tests, which can in turn decrease the population; the resulting absence of contact can, in turn, drive further misunderstanding. More subtly, the very fact of prenatal testing drives the way we think and talk about people with Down syndrome: away from a discussion of citizens with rights and toward a discussion of possibilities, potentialities, and risks. They occupy a kind of limbo of human value. They are discussed—often inaccurately—in terms of their effect on others, rather than in terms of their opportunities and hopes.

In the exhibit hall, I sat next to a young woman, there with her father, who was also selling a book of nonfiction prose. Hers was a slim book of essays, called *The Selected Essays of Sarah Savage Cooley*. I bought a copy of her book, and she bought a copy of mine. On the second day, she said, "Hey, George," and I said "Hey, Sarah," and we chatted about the yoga class she had to go to, which she said she liked. I read her essays, which had originally been published on Facebook. I liked them. Each one was a page or two long, and they were mostly about wanting to be a writer or living with Down syndrome. One, called "What Can I Do," hints that a friend wants her to stop writing:

> I am a writer and a book writer. That is my dream that I am going to keep for the rest of my life. I just can't give up on my dream for someone that I thought I knew who asked me to. He already knows that I can't do that. This is what I need to do for myself and for my heart and dreams: I need to follow my heart and do what's right for myself. I am always the writer that reports the truth about my life and secrets.

She had a much better table display than mine, far more noticeable. She had a stand-up poster board with a large, suitably thoughtful author photo. The book's title and subtitle were printed in large, visible-from-a-distance type, and she had a sample copy available for readers. The sum total of my efforts, in contrast, was to pile books neatly at each corner of my table. After a couple of hours, I hit on the idea of carefully leaning a book against each pile, so that potential readers could see the cover as they approached. Only on the second day did it occur to me to print out a copy of a book review and some blurbs, if only to make the table less spartan. I went all out, paying the FedEx Business kiosk $1.25 per sheet for color copies. I also bought Sharpies, colored construction paper, and tape from the downtown Walgreens. The result, hastily assembled on site, was probably good enough to earn a solid B from Mrs. Tartini, my fourth-grade teacher in 1972; in the elementary school Laura attended, under the current, gradeless dispensation, it would have earned a checkmark somewhere between "Needs Work" and "Meets Expectations."

I've been a parent for many years now, and I feel like an old hand and a newcomer at once. I've reassured young parents (when they seemed to want it) and am reassured myself when I see adults with Down syndrome doing well. At the same time, the Saylor case, for me at least, cast a strange light on my days in Denver. On my second night there, my brain fried from hours of sitting at an exhibit table, rushing to Starbucks for sustenance, and teaching a writing workshop, I wandered the 16th Street Mall. Young women in high heels, a couple of slack-jawed men handcuffed beside a police motorcycle, families ambling the crowded sidewalks stabbing plastic spoons into frozen custard, mimes, pedicabs, an ambulance outside a bar, twins with Down syndrome in a double stroller, the crowd flowing around them: perhaps because I was tired, I felt porous to it all and almost bitter. People seemed to be enjoying themselves, there were lots of people with Down syndrome and it seemed okay, there were diners at Johnny Rockets and Chili's and moviegoers coming out of the United Artists Cinemas, and this was the same America where Ethan Saylor died because he wanted to see *Zero Dark Thirty* for a second time.

I avoided the online comments. Those who did read them told me that they were more vicious than usual. There's no accounting for trolls, and no end to the selective vision and self-deception of outright bigots. Still, I wished that the author of the one comment I did see—that someone like Saylor didn't belong out in public in the first place—could've sat with me at exhibit table #58 for a couple of days. Not that it would've happened, but it wouldn't have been a bad way for someone to discover that people with Down syndrome vary greatly; that they have their own interests; that they are not all happy all the time; that they range, as we all do, in ethnicity, religion, age, body type, interests, and ability; and that they do in fact belong in public, no more or less than anyone else. It is hopeless to argue with a bigot: hatred cannot be refuted, though its distortions can be. But I sometimes think a dose of reality might be useful for those whose hearts have not completely shriveled away.

Many of the fiercest debates around disability center on language: the difference between saying "a Down syndrome child" and "a child with Down syndrome," for instance. These things matter, but not as much as the unspoken assumptions beneath them: our ideas, invisible to us, about invisible populations, and the misconceptions that make our errors seem factual, that lend solidity to the pure wind of our misunderstanding. That misunderstanding exists on a quantum level, easier to deduce than see, but beyond the debates over language are realities hard to ignore, from discrimination in jobs and housing to higher rates of sexual abuse. To say that these result from genetic disability is to commit the old error of favoring shaky biological explanations over obvious social ones, to combine reductionism with blaming the victim.

Which brings us back to the night of Ethan Saylor's death. We now know that Saylor had had an angry outburst outside the theater, and that his aide had called Saylor's mother and another caregiver for advice. They had advised her to wait it out. But Saylor reentered the theater, where the incident occurred. We also know now that Saylor's aide warned the deputies not to touch him, that he would "freak out," and that Saylor's mother was on the way to the theater to help. We also know that Saylor tried four times to dial 411; his mother thinks he may have been trying to get money for the movie.

It is curious to see the way Down syndrome has played into this investigation. Mr. Saylor's condition made him vulnerable, but it was also used to implicitly absolve the deputies of his death. The medical examiner's report "noted that Saylor's weight, Down syndrome and heart disease made him 'more susceptible to sudden death' in situations that compromised his breathing." The sheriff, reacting to news of the independent investigation, also invoked the medical. Interviewed by *The Washington Post*, he made a prediction: "[the investigators are] going to say: 'You know what? There is absolutely no excessive force, no inappropriate actions or wrongdoing by these deputies. This is simply an unfortunate situation where this man had a medical emergency while being escorted out of the theater.'"

In what way did it make a difference that Ethan Saylor had Down syndrome? He was handcuffed, his larynx fractured, with three grown men on top of him: to cite his weight verges on the perverse. It seems far more likely to me that Ethan Saylor died, not because he had Down syndrome, but because of the way that syndrome was perceived. His condition was visible; his personhood was not.

William Carlos Williams, the great poet and physician, once wrote,

> It is difficult
> to get the news from poems
> yet men die every day
> for lack
> of what is found there.

I always loved those lines, but in Ethan Saylor's case, they seem to me literally true. If "what is found there" is the recognition of another intelligence, the compassion for another, and above all *understanding*, in every sense, then in fact Ethan Saylor did die for lack of what is found there: he was not understood, and he was apparently deemed either not understandable, or not worth understanding.

Saylor's case offers a glimpse of a simple injustice. It also shows how slow our culture has been to develop in including those with disabilities. For this reason, the finding that the extra chromosome might one day be "silenced" holds equal shares of trouble and promise. Before we silence a form of expression, we should ask if we have understood it. Before we cure a condition, before we erase it, we should ask if we have seen it clearly.

Sarah Cooley again, from her short essay "Having a Disability":

Having special needs: it feels different. What can I do to make this world understand what I feel? Sometimes I feel that having a disability causes a lot of trouble, and I feel like I am in the way. Sometimes I feel it is my fault…I am trying my hardest to make my life look easy, but it is really hard having this disability, which I know I will have for the rest of my life. But I will keep trying to be the best person I can be.

10 Finding a Place

Now and then Laura writes me a note. In June of 2016, for example:

> Happy father day

> Love you so much Great father Great time with you your gigs awesome your rockstar

> Laura

I am not a rock star. Gigs with my band are mostly in sleepy Northwestern bars with three to five IPAs on tap. The Venn diagram of George and Rock Star looks like two circles, very far apart. Still, I keep the note taped up over my desk, along with the others. When I look at them, I imagine Laura writing in her downstairs room, at the desk which, like mine, is a hollow-core door on sawhorses. She tears the note from the spiral notebook, folds it in half, and scrawls DAD on the outside. Then she hurries up the stairs to find me and thrust the folded note into my hands.

Writing about Laura has its complications. For each sentence, I could add a page of narrative caveats, a fine print to govern the legal interpretation of anecdote:

> This positive description is not intended to inspire. Yes, she is sweet, but also has nonsweet moments. The author stipulates to the existence of said moments but feels no need to describe them for "balance." Despite his numerous positive encounters with persons with Down syndrome, the author explicitly rejects the contention that they share a single, winning "personality" or the underlying assumption that any one can represent all the others. Though the author has strong opinions on a range of social issues, he declines to weaponize his daughter in their service. Laura is not an example in an argument. She is not a success story. She is not a story at all. She is a person, and by describing her, the author intends to suggest what she is like and raise questions about the world she enters. This

work is related to, yet different from, his work as a parent, which is to help her find, in every sense, her place.

Behind these caveats is the wish to control interpretation, and behind that vain hope is, in no particular order, a writer's ego, a father's protectiveness, and a deep familiarity with the average Internet comment section.

Faced with the text of Laura's note, one online reader might see confirmatory evidence that all people with Down syndrome are sweet; another might say, If she's writing, she's unusually high-functioning; another might spew obscenities, abuse, and variations on the word *retard*; another might say, Well, that's all very nice for now, but what happens when she's an adult and the parents die and she's a burden on the rest of us? Saying positive things about a child with Down syndrome on the Internet is like accelerating protons to near light speed in the Large Hadron Collider. You get an explosion of legible trails, and a secondary explosion of analysis and speculation, and in the aftermath, the child seems almost hypothetical: a particle whose theoretical value is much debated, but whose existence has yet to be confirmed.

Complications like this don't much interest Laura. She likes what she likes, which at the moment involves spoiling her rabbit, Skyping with her older sister, refusing help on her homework, and streaming episodes of *Emergency!* on the iPad. Her preference for simplicity, like my fondness for complication, is an effect of identity, of which ability is only a part. We're both working with the brains we have. If she writes *Great time with you your gigs*, it's because I've been playing out a lot lately, and if she puts down things as they occur to her—first thought, best thought—and disdains punctuation, those choices remain appropriate for what a writing teacher might call the Rhetorical Situation. Her style is pure Laura, and for her intended audience, it works.

In one way, Laura's Father's Day note is nothing special: lots of people with Down syndrome read and write. In another, it is evidence of a radical transformation that began in my lifetime and is still underway. In the United States, children with intellectual disabilities only won the right to attend public school in 1975, by way of what is now known as the Individuals with Disabilities Education Act (IDEA). Laura's words express the law's existence, the structures and supports that flow from the law, the teachers and aides who make good on the law's promise, and Laura's own will to communicate. And they express a future, in which her place depends, in part, on her ability to decode the world around her and make her wishes known. But it depends, as well, on what is thought about people with disabilities, and intellectual disabilities in particular.

I wrote, *She is not an example in an argument.* Clearly, though, I have made her one. Perhaps the statement was aspirational. I suppose that as long as the meaning of disability is up in the air, any story will be adduced as an example, whether the writer wants it to be or not. In the case of Down syndrome—recognizable, emblematic, contested—a story is never just a story. And yet to invoke her life, in a book about biotechnology, demands clarification. Her life in the Northwest, our family's life, has little to do, directly, with NIPT or de-extinction, and our one connection to CRISPR is that Theresa uses it in her lab. It's not the technologies that concern me per se; it's the assumptions about people invoked in their support. If Laura is to thrive and have a place, people need to believe that she *has* a place. She has to be perceived, in Allison Carey's phrase, as a "bearer of rights"; if she is devalued, then she is less likely to be so perceived. Whether explicitly or not, the stories that serve new biotechnologies lend force to that devaluation.

Those stories are rooted in a decades-old story of progress, where better genes lead to a better world. The upper-class, disability-free utopias of NIPT marketing; the synthetic cell reframed as a brainteaser for code-savvy intellectuals; the metaphors for directed evolution, in which we, the conscious, intentional ones, improve on Nature, the "sloppy tinkerer," the "deaf, dumb, and blind" maker; the dreams of unimaginably intelligent, posthuman descendants—all have roots in an older story about human improvement, which hinges on an idea of superior intelligence. The progress narrative has a history, and intellect is at its heart.

* * *

In his article "'These Pushful Days': Time and Disability in the Age of Eugenics," Douglas Baynton proposes a surprising argument: that the eugenic conception of intellectual disability resulted, in part, from "a transformation in the meaning of time." Even as evolution had made time seem vast and uncertain, industrial capitalism had accelerated and regularized the days. With labor shifting to factories, many who had once worked in the home no longer had a place. Work was clocked and measured as never before, and the "cultural value placed on speed and efficiency" left people with disabilities out. As a result, writes Baynton, "[o]ften they were described as 'burdens' on the family, as well as on the community and the nation, all of which were engaged in various competitive spheres." Time had accelerated, and in "these pushful days," people with intellectual disabilities could not keep up. This view was reflected in new, time-based words for disability, like *handicapped*—a term associated with horse racing—and *retarded*. Tracing the rise of the latter, Baynton notes that "[e]ducators first began using 'retarded,' sometimes 'grade-retarded,' in the 1890s to describe children who did not advance with their age cohort, for reasons ranging from lack of discipline to malnutrition to mental defect." This came about because of the advent of a "graded school system," whose function "was to instill the values and skills called for in a competitive economy: punctuality, the ability to adhere to a schedule, respect for authority, and an individualistic and internalized work ethic." But the term soon hardened from an adjective to a noun, from a description of a person in a specific context to a reductive label:

> By the 1910s and 1920s…the term had become standardized in phrases such as 'retarded intelligence,' 'the mentally retarded,' and the 'abnormally retarded,' as an umbrella term for intellectual disability. Transformed into a description of a type of person, 'retarded' filled the need for a label that captured the idea of someone who was both uncompetitive economically and a laggard in evolutionary development.

As important as intellect was the powerful idea of "normal." Like intellect, normality is a key thread linking our time to mainline eugenics, and it is essential to narratives of progress, whose cast of characters could be sorted into those who move progress forward and those who hold it back. "Normality on the individual and social level contributed to forward motion," writes Baynton. "Abnormality was a retarding or atavistic force that could potentially slow down or reverse progress."

Intelligence and normalcy remain key ideas for us, their value virtually unquestioned. And yet they elude consistent definition. What counts as "intelligent" varies with time and place. Similarly, the human categories declared "normal" shift from decade to decade. Both are complex, shifting not only in time but also between scientific and common uses. "Normal" is a particularly complex idea, embodying both "average" and "ideal." "From the start," writes Baynton, "'normal' functioned simultaneously as description and prescription…from medicine and sociology to eugenics and education, the normal was also the desirable."

The indeterminacy of these ideas makes them useful. Because we think we know what "smart" means, what "normal" means, the terms are rhetorically powerful ways of joining unlike conditions. "In the early twentieth century," writes Allison Carey, "the diagnosis of intellectual disability simultaneously became more technical and broader and vaguer, allowing for the appearance of scientific objectivity while enabling eugenicists to label many 'unfit' people 'feebleminded.'" Intellectual disability anchored broader categories of the despised and supposedly defective: nonwhite ethnic groups, the poor, and the "sexually deviant," among others. These connections were enabled by the belief that, as Douglas Baynton writes, "[d]efects of the body, intellect, and moral sense were thought to be interconnected." Women in particular were the focus of eugenic control: "the feebleminded woman," writes Carey, was believed to be "without the ability to reason, unable to protect herself, and guided purely by her emotions, uncontrolled sexuality, and animal instinct." Defective, fecund women threatened the country from within; defective immigrants threatened the country from without; for the nation's sake, the borders needed to be guarded, and reproduction had to be policed.

Retardation and *handicapped* are still around, part of a long etymology of labels, a society trying to pin down a concept both critical and elusive. To define it was to define ourselves. James Trent writes that each generation reinvents the feeble mind: "the pitiable, but potentially productive, antebellum idiot and the burdensome imbecile of the post-Civil War years gave way to the menacing and increasingly well-known defective of the teens"; burden, then menace, gave way to the eternal child, the ideal—still contested, still unfulfilled—of citizen. The irony is that the category by which we define ourselves is itself far from definite. There is, as Allison Carey writes, "no clear or consistent demarcation" between people who are and are not intellectually disabled: "There is no 'typical' citizen against whom people with intellectual disabilities can be compared, nor is there some 'typical' person with an intellectual disability whose characteristics can be assessed."

In 1985, the historian Daniel Kevles wrote that the heart of eugenics—in all its forms, whether market-based or coerced—is the idea of "abstraction": "[E]ugenics has proved itself historically to have been often a cruel and always a problematic faith, not least because it has elevated abstractions—the 'race,' the 'population,' and more recently the 'gene pool'—above the rights and needs of individuals and their families." But those elevated abstractions also implied their opposites. The positive qualities eugenicists sought to spread through the population—health, normality, and superior intellect—implied the negative qualities of disease, abnormality, and intellectual deficit. Those negative qualities dominated the descriptions of persons with intellectual disabilities. The people were replaced by abstractions: their intellects reduced to IQ scores, their head shapes noted and compared, their common qualities distilled and iterated in lists, their names replaced by labels. The accounting constituted an erasure, but with the fading of mainline eugenics after the Second World War, a different kind of account would begin to emerge, a story in which people with intellectual disabilities had a meaningful place.

* * *

There are dozens of memoirs about raising children with Down syndrome, hundreds of blogs, a galaxy of status updates. But in the beginning was *Angel Unaware*. Written by Dale Evans Rogers—writer, actor, wife of Roy Rogers—it's the story of Robin, their daughter, born in 1950 with what was still called *mongolism*. Robin died at the age of two, with an unrepaired heart defect, from mumps encephalitis. Published in 1953, *Angel Unaware* both signaled and influenced a turn in thinking about children with intellectual disabilities. By claiming a child who might previously have been seen as shameful, Rogers helped launch the parents' movement of the 1950s, which led, in part, to the rights that Laura enjoys now. The book occupies the extremes of familiarity and strangeness: written before the advent of prenatal diagnosis, the disability rights movement, and open heart surgery, *Angel* engages enduring questions in a lost context. How is this person to be imagined? What does it mean to care for her? What is her place—in the home, in the family, in the world? And why tell her story?

It has to be said: to a secular reader in 2017, *Angel Unaware* is a spectacularly weird book. It is written in the first person, with Robin as narrator. As Rogers explains in the Foreword, "This is Robin's story. This is what I, her mother, believe she told our Heavenly Father shortly after eight p.m. on August 24, 1952." The book asserts that Robin was "a tiny messenger," sent by God "on a two-year mission to our household." Rogers is a kind of New Journalist of heaven, offering an imaginative reconstruction of a divine interaction. Her book was intended (and received) as an inspirational memoir, but in genre terms, the book is a hybrid of science fiction, Westerns, sermon, and reporting from the Beyond.

Angel Unaware tries to depict a stable world: one in which God has a plan, suffering has a purpose, Heaven is for real, and the meaning of experience is clear. But reality keeps breaking through, and so, in practice, the narrative projects ambivalence, uncertainty, and unresolved contradiction. The book's central conceit, for example, treats heaven as fact, time- and date-stamping Robin's words from the eternal. And yet when Rogers writes, "This is what I, her mother, believe she told our Heavenly Father," the word "believe" wavers between reportage and invention. It implies knowledge of a literal heaven while highlighting the mother's inability to know for sure.

Rogers's device also offers an early example of a parent resisting a purely medical narrative. By giving the exact time and date of Robin's words, the sentence transforms the monotone of medical record ("the patient died shortly after eight p.m ...") into a transcendent rebirth. Death is a new beginning, a deeply Christian idea that is mirrored by the book's form: Robin's death occurs in the Foreword, prefiguring her rebirth as text, as a message and a voice.

And yet, for all its inspiration, the book betrays a deep ambivalence about Robin herself. Rogers can only assert her daughter's value by erasing her, can only write her by overwriting her. Angel Robin is idealized: sweet, thoughtful, childlike, intelligent, wise. She is naive about history: "I wondered what Mongoloid meant. They seemed to think it was something awful." She is Christlike, a child that redeems, a divine human on an earthly mission. And yet Actual Robin and Angel Robin coexist side by side, unreconciled. They are juxtaposed in the title—"angel" describes the heavenly Robin, "unaware" the earthly one—and the juxtaposition is even clearer in Angel Robin's memories of language and development: "I had eight big teeth and I could chew crackers, which I called 'cack-cack'." A nurse is named *only* by Robin-as-Human-Baby: "Cau-Cau." Her inability to speak is couched in fluent sentences; disability is nested in ability.

Though Rogers was clearly devoted to her child, the book manifests a curious lack of connection between mother and daughter. The narrative suggests care, more than love. Robin is an agent of good, and her truest bond is with her handler, her Heavenly Father, the unseen Charlie to her angel. Robin addresses Him with a conspiratorial "we," referring to the desolation below either with serene distance or a chirpy optimism. And yet the book bridges the very distance it creates: the narrative voice constructs a deep togetherness, a communion where mother and child exist, throughout, in a single line. That communion reads as both wishful and elegiac. Rogers finds, if only in narrative, an answer to the question faced by parents: What connection can there be between an able parent and an intellectually disabled child?

In this way, too, the tidiness of the book's universe is betrayed by the narrative. The book portrays a loving, traditional family (if a celebrity one, with performing commitments), its existence overseen by an omniscient deity. But throughout the book, identities multiply and blur, and lives are filtered through personae. Robin is speaking, but so is her mother, who is always present (as the author), and also present, in flashes, as a character. The voice of Robin often sounds like a folksy, optimistic character from a stock Western, as if Dale Evans, the movie persona, had been absorbed into the book's angelic voice. (Referring to "her" difficulties with speech, Robin says, in her Blessed Infant Cowboy Talk, "My balky old tongue wouldn't behave.") Robin, an angelic fiction, is composed of other fictions.

Perhaps that fiction allowed Rogers to reveal herself more fully. Through Robin's voice, we see confessional glimpses of the author, who is by turns desperate, possessive, guilty. Because of modesty, Western stoicism, or the absence of a tell-all tradition of memoir, we don't see much, but what we do see is moving. For Rogers, "care" means not only the work of raising a child through medical crises and developmental delays, but also "worry." Robin is on Earth to teach lessons, but also to test her parents; despite the optimistic worldview, it is clear that the test was severe. (In the Foreword, Rogers refers to Down syndrome as "an appalling handicap.") *Angel Unaware* is an introspective book by someone who does not especially believe in introspection, and as a result it feels more exposed than most. The writing has the chatty tone of a precocious toddler, but it pours out with the force of a sob.

At the same time, Rogers's strategy of fictionalizing her life, of filtering it through characters, is itself revealing. If *Angel Unaware* treats identity as blurry, manipulable, if it invents Robin in order to represent her, that strategy is perfectly consonant with Dale Rogers's life, which had been a series of reinventions. She had been born Frances Octavia Smith. "Dale Evans" was a chosen stage name—her third, after "Frances Fox" and "Marian Lee." Roy Rogers (also a stage name) was her fourth husband. She had eloped at fourteen and given birth to a son at fifteen, whom her parents helped her raise while she pursued a movie career. (For a time, 20th Century Fox pushed the fiction that her teenage son was really her younger brother.) With Roy, her main business was the manufacture and display of personae. As her biographer notes, she and Roy starred in movies and TV shows, all of which featured tidy narratives of good and evil set in a mythical West. (She also wrote the song "Happy Trails," which has seemed more melancholy to me ever since reading *Angel Unaware*.)

These strands—family, religion, performance, disability, identity—are woven together in a pivotal scene near the book's end, one which distills the book's contradictions. In it, Roy Rogers is singing "Peace in the Valley" to an audience in Houston. It is shortly after Robin's christening. Robin was not present for the performance, but she gives voice to her mother's memory: that is, Dale Rogers imagines Robin-in-Heaven imagining Roy Rogers's performance and commenting on the lyrics, addressing her Heavenly Father about her earthly Daddy:

> There'll be peace in the Valley for me,
> … peace in the Valley for me.

I wonder if Daddy is thinking of *his* peace, or mine, when he sings that?

> There's no sadness, no sorrow, no trouble I see!
> There'll be peace in the Valley for me!

No sadness, no sorrow! No crippled children, Father!

> There the bear will be gentle and the wolf will be tame
> And the lion will lie down with the lamb;
> There the host from the wild will be led by a child …

A child like *me*, Father?

> And I'll be changed from this creature that I am.

Given Robin's story, the long history of comparing people with intellectual disabilities to animals, and the lyrics about bears, wolves, lions, and lambs, the phrase *this creature that I am* has a complex resonance. In the case of Robin, who is limited to syllables ("cack-cack," "Cau-Cau," and so on), the divide between the earthly and heavenly selves is widened. Robin's human sounds are the syllables of a "creature"; her narrative voice belongs to an angel. In the terms of an older framework, she has ascended the Great Chain of Being. In contemporary terms, she has been enhanced. She has wings; her intellect has been upgraded; she has the power of speech.

Here, too, the view of Robin is deeply conflicted. She is, on the one hand, a symbol of ultimate peace: "a child like me" is associated with an Edenic state. At the same time, she is made to voice the sentiment that children like her would be better off not existing: "No crippled children, Father!" Further, she is valued only by being transformed: "I'll be changed from this creature that I am." In other words, Robin is only valuable if she is either improved or erased. Both alternatives suggest that the child is unacceptable as she is. To say so is not to indict the book, or Dale Rogers herself, who criticizes a doctor who suggests that children like Robin should be "machine-gunned." *Angel Unaware*, in its flawed way, attempts to speak for Robin's value, but it is also a book of its time.

More than sixty years later, we are still contending with narratives of improvement and erasure, but changes in technology and culture have complicated the questions Dale Rogers faced. The availability and routinization of prenatal screening, and the cultural and legal gains associated with the disability rights movement, have amplified contradiction, not erased it. Ironically, though, Rogers's perspective on her daughter may be helpful here. Absent a genetic explanation, she joins Robin to other children who are "handicapped," *no matter the cause*. If her work is, as literature, accidentally experimental, her politics are accidentally radical. By taking a child usually considered subhuman, and reframing her as superhuman, she made her *more* valuable than other children. She inverted the system of value by which "Mongoloids" were despised.

Like today's parents, Rogers faced a disruptive event that set her at odds with the dominant perception of disability; like today's parents, Rogers felt compelled to respond with imagination, to craft a moral narrative. That narrative is deeply flawed: by erasing and upgrading its subject, by transforming the real child into an angel, it embodies the very ideas it tries to resist. And yet, in an age of increasing prediction and control, it is worth remembering that Rogers, above all, spoke up. The contemporary accounts I value most are unambivalent about a child's value: they are stories about citizens, not angels. But a story with an angel in it is better than no story at all.

In 1953, the same year *Angel Unaware* was published, James Watson and Francis Crick discovered the structure of DNA. Six years later, Jérôme Lejeune would be credited with discovering the extra chromosome associated with Down syndrome. Soon would come amniocentesis and legal abortion, radically altering the story we tell about Down syndrome, even as the institutions began to close and children won the legal right to enter school. The interlacing of biology and computers would bring the Human Genome Project, NIPT, CRISPR. New eras would be declared, though each new era overlapped with the previous one or altered it. Though it is possible to list developments in order, like layers of sedimentary rock, each new invention is more like an earthquake or volcanic intrusion that radically alters the landscape, leaving it both familiar and disarranged. To tell a story about Down syndrome, in 1953 or 2019, is to be aware that the date does and doesn't matter, that the landscape is a shattered mix of ideas and practices both new and old. In one sense, we move forward: the progress we take for granted is real. In another, the past isn't even past.

*   *   *

To know and love someone with an intellectual disability is to question what you thought you knew. Chris Gabbard, whose son August had cerebral palsy, writes,

> I grew up prizing intellectual aptitude…and detesting "poor mental function." Perhaps what helped make me revere intelligence was growing up in Palo Alto, with Stanford less than half a mile away and a number of Nobel Prize winners and famous and wealthy technology innovators all around me. People in my immediate vicinity had good brains, and that meant money, respect, and international influence.

After August was born, Gabbard, a writer and English professor, found himself less certain about the importance of intellect and more interested in questions of belonging. He asks, "Is it really true that the unexamined life is not worth living? And is it accurate to say that only the possession of *logos* qualifies an entity for human status?" He writes, "the…Augusts of the world are as much members of our human tribe as any of us are," and then asks, "in an academic environment that rewards being smart, how do I broach the idea that people with intellectual disabilities are fully equal?"

In competing narratives of American progress—one aimed at belonging, the other at improvement—intellectual disability is a key character, but in one it is the quality to include and in the other it is the quality to erase. This is a problem of power, and it's compounded by the fact that people with cognitive disabilities often find it difficult to respond. When they can, our job is to listen. When they cannot, the question is who will speak for them, and how. Parents are often the ones to speak, but this enterprise is fraught with risk. Now that Laura is a teenager, I can ask her if it's okay with her to tell a story about her; if she's not sure, I don't. When she was younger, though, I went on faith. There is, as I regularly tell students, no algorithm.

When parents and others do speak about or for people with intellectual disabilities, we need to do so responsibly. For a start, that means surrendering rhetorical techniques that are crude but effective. It's easy, for example, to play to positive stereotypes of Down syndrome, but doing so would buy into everything I reject: Taking the individual for the group. Accepting the medical classification of that group as the primary one. Auditioning for acceptance, making the child a reality show contestant, canned story and all. A victory on these terms is not worth having, because it leaves the larger conceptual errors untouched, and how we conceive of people with disabilities is crucial—which is to say, how we conceive of *we*, of who is counted in the first person plural.

Highlighting positive qualities may also buy into the belief that people with intellectual disabilities have to have special qualities in the first place: that if they fail at their duty to contribute, they then have a duty to inspire. But this belief suggests that they matter because they affect us in one way or another, and not because they are citizens with interests of their own. Even the best-intentioned writing can obscure its subject. Debates over how to represent a child risk falling into the trap of believing that the child has to mean something in the first place, which in turn reproduces the core problem of focusing on What the Disabled Mean to Us, which in turn excludes Them from Us, inscribing the very separation we should try to erase.

And story has intrinsic limits. First, a story can be falsely representative: a person depicted can be taken as standing for the group or encrusted with meanings the person might not choose. Second, storytelling ability may be taken as evidence of human value—*I narrate, therefore I am*—an assumption which embodies the devaluation of many people with disabilities, as not everyone has the ability to tell or understand stories. Third, there's nothing inherently good, in a moral sense, about stories, which can embody mistaken or toxic ideas about people. And last, we live in a digital environment now, of which stories are only a part. Most of our entertainments are synthetic, combining images (still and moving) with words; most are interactive in a different way than stories, though not necessarily better.

As a writer, I think about *the* future. As a parent, I think about *Laura's* future. Lately, that future seems more fragile. That's partly because it's nearer than it used to be. Parents call it "the cliff": the moment, at 21, when a child ages out of the school system. What's left, after you go over the cliff, are few to no options for work, housing, and an independent daily life. It's common to point out that intellectually disabled adults "fall through the cracks," but this metaphor is not quite right, because it suggests a landscape that is mostly solid. It is mostly made of cracks. Most adults with intellectual and developmental disabilities don't work, even though most want to, and fewer still receive a living wage. Most don't live independently, even when they could, and many face discrimination, exploitation, and abuse. That this is true, and that it is also true that Laura loves her high school, suggests the extremes of contradiction we face.

Laura's a good kid—she works hard in school, people love her, and she rarely gets lunch detention, and we're incredibly proud of her for all that—but I don't want her successes used to suggest that everything is peachy, educationwise, in the United States (it's not), or turned against others who have a harder time with achievement or behavior. Some act out. Some students are nonverbal—though speech, it needs to be noted, is only one way to communicate. Many people with cognitive differences are not intellectually disabled at all, and "intellectual disability," though tonally superior to the outmoded "retardation," tends to collapse a wide and extremely varied category. The point of a story is not to gloss over the differences or the very real difficulties; it's to ask what would be required for everyone to have a valued place.

* * *

In 1967, the first autobiography of a person with Down syndrome was published: *The World of Nigel Hunt*. The book has a long preface by Nigel's father, Douglas, and a shorter preface by the scientist Lionel Penrose. In his preface, Douglas sounds like a man describing the paranormal to a skeptic. He recounts the disbelief, and conversion, of teachers and experts. He cites Penrose, then the preeminent expert on the syndrome, who notes that Nigel is a "high-grade mongoloid" but is careful to point out that he is not a mosaic. Douglas even includes a facsimile of one of Nigel's typed pages, to rebut the accusation that Nigel did not really write the book. (Mosaic Down syndrome is a rare condition, in which some cells have 46 chromosomes, and others have 47. People with mosaic Down syndrome are less severely affected. The fixation of skeptics on mosaicism is profoundly irritating: Traveling to read from *The Shape of the Eye*, I'd now and then be asked, after recounting some small success, if Laura was "a mosaic." The assumptions are troubling: (a) that any achievement must have a biological basis, and have little to do with schooling, parenting, community support, or the student's own efforts; and (b) that any achievement that seems exceptional *to the hearer*, who invariably has low expectations, can only be explained chromosomally.)

It is difficult to imagine *The World of Nigel Hunt* being written by the sober adult of the preface. The book has a thinking-aloud quality; Nigel corrects himself as he goes. ("Since President Kennedy's death at Dallas, Texas I was most cross, well, extremely sorry.") He addresses himself directly to the reader in the introduction and seems to be thinking of the reader as he writes. On the death of Winston Churchill: "Do you remember him where he sat with President Eisenhower and the other president? I do too." He is aware that, in telling a story, he is creating a picture in the reader's mind. He writes down what he remembers, seemingly as he remembers it, but he is also a docent of his memories. "After we finished looking at the Air Force Memorial I would like you to go to Stoke Poges where Thomas Grey the poet was buried." He is associative, not strictly linear; he is reminded of other days and writes them down, but he is aware of switching from one story to the next. "Sorry I can't tell you any more of Stoke Poges," he writes.

But most of all, Nigel likes to look at things: the book is a long, associative travelogue. He notices small lizards on the wall of a church. Describing the "Blau See," he writes, "It is very blue and we could see a lot of trout floating around and you can eat them." He assumes that his father does the backstroke in order to see things too: "In Italy my father swam a lot, he can even swim on his back so he can look up at the mountains, and a good job too." A couple of pages later, Nigel describes "the swimming pool where father swims on his back in the cool air of Zirl. Still on his back he sees mountains and the lovely clouds." Eight pages later: "My father got to the edge [of the lake] and like a trout he went in. Mum and I were sitting on the right side of the pool, and he swam on his back. He saw four eagles flying." Just as Nigel imagines the reader's perspective, so he imagines his father's. (Douglas notes, perhaps with a touch of impatience: *Being well into my sixties and having functioned on one lung for thirty years I am quite incapable of swimming other than on my back.*) It is a book where the parent lives in italics, in brackets—where the parent is both witnessed and witnessing—and yet, as is often the case, the parent is there to explain, sometimes defiant, sometimes almost apologetic. There is a mile-wide gulf between Dale Rogers's ventriloquism of her late daughter, and Douglas's occasional editorial interjections, but it is true that in most cases, parents will be more prominent as interlocutors for their children. Speaking about, speaking for, adding context, interpreting. It is the interplay between Douglas's voice and Nigel's that charms me because it resonates with my own life.

In two ways, Nigel Hunt's writing passes Creative Writing 101 with flying colors: the "voice" is distinctive, and the writing is vivid. But reading his book helps me see that these criteria are also about in-groups and out-groups, about intellect and status. A *distinctive voice* stands for *someone worth listening to*, a status achieved by the performance of intellect: wit, metaphor, surprise. And the exhortation to be vivid—"show, don't tell," in the moldy formula—is really a proxy for intellect's value. Valuing concreteness *is* valuing abstraction. It's not "Put in specifically colored wheelbarrows, which will make your poem more appealing." It's more like, "Using sensory details will encode an abstract concept and/or feeling for the reader, plus identify you and the reader as fellow In-Group Members who understand the importance of meaning-charged sensory details." Implicit in the same advice is the value of implication, which itself encodes intellect's value: you don't want to "insult your reader's intelligence" by "spelling things out." This message, of course, is delivered overwhelmingly in university classrooms, where people with intellectual disabilities are rarely present—though this, too, is changing.

Reading *The World of Nigel Hunt*, I wonder, against all my carefully thought-out reasons, if there is a profile, a common thread of personality, associated with the condition. It is far more complex than it has been made out to be, and it is a starting point, a set of possibilities; but I've often been struck, seeing one of Laura's friends, by indefinable similarities, a core of resemblance. In the tilt of a head, in the throwing of a ball, in a sound that is not a word yet, but becoming a word—that differs, from age-matched language, as syllable from sound—there *is* something there, as if the chromosome had been struck like a church bell, and the clustered overtones were fading across the grass. The residue of an unheard music.

Nigel seems perfectly happy with his life. Not *genetically* happy, as is believed, not automatically sweet, but happy to get to travel with his father. But nonfiction is always partial. What the reader sees—and, reading, reinvents, rebuilds, with the tools of her own experience—is a slice, a performance, a transformation, a Steadicam shot tracking a single point of view. In the course of the writing, Nigel's mother dies. The voice does not change much, but as Douglas writes, the event was confusing for Nigel, resulting in a breakdown not narrated in the book.

It was years before the field of disability studies got going, but the book illustrates what might later be called interdependence: Nigel's web of relation with his parents; the skills he had, because they taught him, believing it possible for him to learn. At the same time, Douglas's italicized narration offers context for things Nigel didn't understand or wasn't interested in. Like many parents, and more recently like many with intellectual disabilities, Douglas offered Nigel's story as evidence in opposition to a restrictive medical view. He does so with a scene:

> Before Nigel was five, I was summoned to the senior officer concerned with mental affairs (in a certain county). My wife and Nigel and I went to see this "expert," who was to help decide our child's fate.
>
> The first thing the good lady said to us—in Nigel's hearing, of course—was, "Oh, yes, a little mongoloid. Quite ineducable. Do you want him put away?"
>
> Had we been more easily impressed by "experts," we might have said "Yes."

Conclusion: Bodies and Homes

Seventeen years ago I drove out to the Oregon desert with a friend and saw the aurora borealis. I was at the beginning of a story that is still unfolding. Laura's arrival was seismic, a subduction event in my mental landscape, and along with everything else, what was upended was the system of human categories by which I understood the world. I thought I needed to understand Down syndrome better, which was true—I was totally ignorant—but this was really the least of it. I needed to think about *people* in a different way, so that I could help Laura toward an everyday belonging.

If I open the Finder on my laptop and navigate to *Macintosh HD> Home>G—writing>Shape—Dead Files*, I find that **auroraborealis.doc** was last saved in 2002. Fourteen years later I wrote the last three paragraphs, saved the essay as **auroraborealis—final.doc**, and published it online. Despite the filename, nothing is ever final, especially in understanding disability, and my shifting understandings have a digital location, a folder called *home*.

The ordinary language of digital housekeeping—*file, folder, home*—is metaphorical. As with DNA, our metaphors traverse space and time. We use something human scaled and graspable, like *book of life*, to imagine a genome a few atoms wide and a yard long; we use metaphors from the last age to imagine the one we live in now. A file in a folder on a desktop. A cell as a factory. The tendency of biotechnology is to superimpose one time on another: de-extinction superimposes a distant past on an imagined future, prenatal tests give the near future weight and shape, and CRISPR allows us to leverage present values to create the future, to write our ideas about life into being. The story told to advance these technologies mimics their effect, calling on the past's solidity to lend shape to the future. Its present is haunted by other times.

This book was written on computer, about half the time. The rest of the time, I wrote longhand. The computer began as a convenience but soon became a necessity, and then something like a cog in thinking. I write with less silence than I used to, and I spend less time with a pencil above a blank sheet of paper, more time staring at a screen, doing keyword searches of the vast, disorganized piles of text that the computer allows me to access and accumulate. I was only thinking of the right word, and all along I was drifting with the digital current. In the same way, I wasn't on Facebook, and then I was; I didn't have a cell phone, and then I did; I didn't text, and then I did; I didn't have a smartphone, and then I did. Our lives are pervaded by the digital, the thread of each individual story is woven into a warp of zeroes and a weft of ones, and mine is no exception. I am in between an old way of doing things and a new way, as many of us are. But the species, too, is in between an old way of doing things and a new way, between the past of Ordinary Human Reproduction and the digitally assisted forms now underway.

* * *

In July of 2017, a colony of bacteria was made to store a short movie for playback, like a living DVR. In a paper from George Church's lab, with Seth Shipman as first author, an iconic few frames of early film—Eadweard Muybridge's movie of a galloping horse—were turned into a GIF, whose digital code (zeroes and ones) was recoded as a genetic sequence (As, Ts, Gs, and Cs); that sequence was used to construct short stretches of custom DNA, which, using CRISPR-Cas9, were then embedded in the genomes of living cells. By sequencing the cells, Shipman and his team were able to accurately reconstruct the movie, in order. You can watch it online: a few seconds of the mare Annie G., filmed galloping, in 1872. The pixels are huge, like in Pong. But watching the movie now, in 2018, splices multiple times together: it feels like the dawn of moving images, the dawn of the digital age, and the dawn of the fusion of biology and computers, all at once.

The same paper also reports a preliminary result: encoding the image of a human hand. Both the still image and the moving one were consciously chosen. Interviewed about the project, Shipman said of Muybridge's movie, "It was one of the first examples of a moving image, it was captured with a technology that was very new at the time, and it answered some relevant questions." As for the hand, "[i]t harkens back to the first images that humankind put in the natural world—handprints on cave walls," he says. "We're putting images into the natural world in a different way."

The choice of this moving picture suggests a conscious act of revision; in a way, it de-extincts Annie G., bringing the animal back in a different form. The image is nominally the same but conceptually different: Muybridge turned a living horse into a movie; Shipman revived the movie in a living cell. Like Craig Venter's Synthia, the project exemplifies the rapid, category-disrupting transformations made possible by the merging of computers and biology. NIPT moves from blood to hard drive to paper; the GIF of a horse moves from film to hard drive to cell to hard drive to screen. That I was able to print a PDF of Shipman's paper at home reminds me that each transformation, blood to paper, movie to cell, is only the most visible part of unending flux, that each transformation sets others in motion, and that these changes reboot the relation of life to time. Extinct creatures return, Muybridge's movie of a horse—the very first blockbuster—is rebooted in a petri dish, and a new story is written with six-letter DNA. As if life were a newly complex equation, and newly unknown.

In her book *River of Shadows: Eadweard Muybridge and the Technological Wild West*, Rebecca Solnit contends that the technologies that arose with industrial capitalism altered the human understanding of time. A telegraph meant that messages were instantaneous. Railroads turned a months-long cross-continental journey into days. Photography conferred the ability to freeze time and save it. Motion pictures allowed time to be slowed, sped up, reversed, and dissected. The very technologies that signified linear progress also undercut the linear. In Muybridge, Solnit sees a forerunner of our present day: Just as Muybridge, filming a lone horse on a racetrack, could "seize hold of that running, stop it on film, take it apart and put it back together," so today's technologists can fragment and reassemble life itself. Muybridge's work led to both Hollywood and Silicon Valley, and Silicon Valley's reach now extends to the gene: we move "from the nineteenth-century transmission of people and materials to the present-day transmission and manipulation of the microcosmic, of electrons and genes."

The persuasive story told about biotechnology is haunted by two understandings of time. One celebrates the linear: a progress narrative, in which humans (aided by science) accelerate away from vitalist superstition, in which past industrial revolutions lead to the information age, and randomly driven natural selection yields to the age of consciously directed evolution. But the same narrative celebrates disruption: linear progress is evidenced by the power to disrupt the linear. Beneath the promise is a paradox.

That paradox informs the persuasion attached to the technology. The happy families in NIPT ads, for example, are a promissory vision of the near future; the images of beautiful babies, of serene mothers late in pregnancy, make that desired future real, superimposing it on the present. Another example, from the field of de-extinction, is the panoramic image of living and extinct species at reviverestore.org. It shows silhouettes, some living, some dead, all occupying a sort of savannah; beneath them is a timeline, beginning in the present, extending back through the history of vertebrate life. In the image, all time is visible in a single picture, as if the mammoth or pigeon could simply move across the image to the living present. As if they were already here. One reason online media work so well in the pitch for new biotech is that, unlike words, they are freed from the tyranny of the linear: able to combine words, pictures, and video, they imitate the disruption they favor.

Biotechnology both disrupts and emphasizes linear time, but so does disability itself. In *Feminist, Queer, Crip*, Alison Kafer notes that the medical terminology for disability and illness is temporal—*chronic, constant, intermittent, prognosis, developmental*—and that progress is imagined in terms of a future without disability: past eugenic practices were "justified by concerns about 'the future' and particularly future children," practices which extend beyond disability to other marginalized groups. She asserts the need for writing that responds to the dominant narrative. "To put it bluntly, I, we, need to imagine crip futures because disabled people are continually being written out of the future, rendered as the sign of the future no one wants." She argues for "futures that embrace disabled people, futures that imagine disability differently, futures that support multiple ways of being."

* * *

In 1973, when I was eight, I left the nurturing, gradeless, unstructured private Montessori school my mother still blames for the way I turned out and entered the public elementary school in Bergenfield, New Jersey. I learned the boredom of watching the clock. I had "learned at my own pace" at the private school, so I was ahead of others in the class, a fact of which I was inordinately proud.

We had never gotten report cards in the private school, but we received them in the public school. I took to these easily: they were a language I understood, a simple five-letter alphabet of worth. Today, these would come home directly to the parents, but then it was heavy beige card stock, with grades written in red pen, one for each subject, neatly inscribed in its rectangle. I remember seeing a girl's report card with failing marks. Did she share it with me? Did I glance across a desk, furtively? I have a vague memory of the girl's face and a precise memory of my shock. I remember that such a failure was inconceivable, and that my personal horror of getting less than an A transformed my understanding of the girl, whose name I have forgotten. I felt less contempt—across the gulf from A to F, what contempt was necessary?—than wonder. At that age, I was aware that having a Japanese mother and a Jewish father made me different, and the occasional epithets *chink* and *kike* drove the point home, so I had a dim but real sense of the importance of human categories. This did not extend to ability, where my eight-year-old's sense of human groups was as rigid as Francis Galton's. To be an A person was not to be a B person, let alone an F person. I did not question the value of being smart; my identity depended on it, though I believed it to be a sailboat I was steering capably through the waves, and not a spar in the wreckage I clung to, trying not to drown.

The Education for All Handicapped Children Act—later renamed the Individuals with Disabilities in Education Act, or IDEA, the law on which we and Laura depend—was two years away from passing. In the more than forty years since, everything has changed and nothing has. I have the sense of eons passing since then, and yet, reading the casual contempt for intellectual disability in the words of bioethicists or comedians, I see that time stands still. Their words are poisonous madeleines, calling me back to something I used to believe, or that still lives in me, contained, surrounded.

*   *   *

How we treat people depends on what we think about them, and what we think is both revealed and influenced by what we say. Which is why the rhetoric attached to biotechnology—the rhetoric of the makers—matters. "To become a maker," writes Rebecca Solnit in *The Faraway Nearby,* "is to make the world for others, not only the material world but the world of ideas that rules over the material world, the dreams we dream and inhabit together." We have dreamed of separation and built actual brick walls to match. We built institutions because we believed the people did not belong with us; we sterilized them because we wanted fewer of them, because we believed that "feeble-mindedness" and "criminality" were linked; we failed to educate them because we thought them incapable of learning. People with disabilities have lived in the world of our ideas, and in the world built by our ideas. They have not, until recently, been consulted; their dreams have not been noted; they have been deployed as examples and cautionary tales, but not often recognized as persons; and so we have not inhabited the world together.

The history of intellectual disability in our country is mostly tragic, though it is tragic in the colloquial, not the literary sense. *Tragedy* has an arc—a hero, brought down by hubris, to self-knowledge. The hero is a great man who falls, and the victims of eugenics never rose in the first place. They had no status to lose. Reduced to timeless categories, they could therefore have no stories. So many of their lives were foreclosed, wasted, suspended: those stories are largely untold, because stories are social, and to have a story you have to interact with others, to have the possibility of change, to have enough status to be recognized. Lives were frozen in the stasis of institutions, and bodies were contained from without and violated within by legally sanctioned sterilization. They did not choose their homes, and their bodies, the last sanctuaries, were violated too.

Those tragedies were made possible by invisibility, and the invisibility was made possible by segregation, shame, and silence. People with intellectual disabilities flicker like ghosts through history, disappearing from view, reappearing in the public mind after a scandal, and fading from sight again. That my daughter's condition is called *Down syndrome* owes itself to such a scandal: John Langdon Down took over the Royal Earlswood Asylum for Idiots after a young woman's toes had to be amputated from gangrene, and it was there he would discover the people who bear his name. The same pattern holds from then to now: scandal, publicity, reform, backsliding, scandal. Someone would die from burns. Someone would stumble on a regime of beatings and solitary confinement. A reporter would film patients sitting in their own excrement. The late 1960s would see a wave of deinstitutionalization, but not an end to the stories, which sound the same from decade to decade.

Here is the reformer Dorothea Dix, describing an institution in 1850:

> ... there have been hundreds, nay, rather thousands, bound with galling chains, bowed beneath fetters and heavy iron balls, attached to drag-chains, lacerated with ropes, scourged with rods, and terrified beneath storms of profane execrations and cruel blows; now subject to gibes, and scorn, and torturing tricks—now abandoned to the most loathsome necessities, or subject to the vilest and most outrageous violations. These are strong terms, but language fails to convey the astounding truths.

Eunice Kennedy Shriver, in 1962:

> I remember well one state institution we visited several years ago. There was an overpowering smell of urine from clothes and from the floors. I remember the retarded patients with nothing to do, standing, staring, grotesque-like misshaped statues. I recall other institutions where several thousand adults and children were housed in bleak, overcrowded wards of 100 or more, living out their lives on a dead-end street, unloved, unwanted, some of them strapped in chairs like criminals. In the words of one expert, such unfortunate people are "sitting around in witless circles in mediaeval prisons." This is all the more shocking because it is so unnecessary. Yet institutions such as these still exist.

Michael Dukakis, remembering a visit to the Fernald State School in 1967:

> "We were in a building with a concrete floor, kind of the size of a small basketball court," he recalled. "There were 75 boys, half of them naked, sitting in their own excrement. The stench was overpowering. They were rocking back and forth for hours. With two attendants. In Massachusetts. Unbelievable."

From an article on Washington State institutions in the Tacoma *News-Tribune*, 2017:

> In November of 2016, a staff member at the Rainier School sexually assaulted a female resident. An investigation revealed later that several other residents had allegedly been raped by the same staff member. The accused awaits trial in the Pierce County Jail.
>
> Employees at Rainier School said training on how to identify sexual trauma in nonverbal adults was never administered.
>
> At that same institution, in the span of less than two years, two residents choked to death, and a man nearly drowned during a lake trip. A staff member left him alone on a dock strapped into his wheelchair. When he fell into the water, he was unable to free himself.
>
> The report also detailed how staff at the Lakeland Village facility near Spokane withheld food to "manage" behavior. The plan called for staff to "only provide diet supplements" if they saw the resident come out of his room.

Two motifs recur in these reports. The first is shock, expressed by witnesses from 1850 to the present: Dorothea Dix notes "the astounding truths," Eunice Kennedy Shriver describes conditions as "shocking," and Michael Dukakis describes his memory as "unbelievable." The second is the distortion of time: the witnesses are shocked that the abuses could occur in a present they believed to be basically just, that "institutions such as these," in Shriver's words—ones likened to "mediaeval prisons"—"still exist." It was "unbelievable," in Dukakis' words, that they could exist "in

Massachusetts." In the same article, Dukakis tours the New England Center for People with Autism with a reporter; it's described as clean, welcoming, positive. "We've come centuries," he says. The article on the Rainier School asserts that the reports on the school's abuses "read like documents from the 19th century when people with Down syndrome, autism, or other disabilities were isolated in asylums and often neglected."

Beneath the mix of shock and time is a belief in progress. But when the conditions that lead to the suffering are the same—the stigma attached to people with intellectual disabilities, the absence of a community place for them, the defunding of programs meant to serve them, and the operation of programs "away from the public gaze"—then we should not be surprised when the misery repeats itself. The individual cruelties of staff members notwithstanding, the living conditions were abject and horrific but not intended as such. No one set out to create an institution whose *purpose* was to subtract rights and inflict pain. It happened naturally, organically, as a result of neglect and indifference and invisibility, which were rooted in a commonsense devaluation. Sometimes that devaluation was rationalized in moral terms ("they" were criminals or sexually deviant), sometimes in economic terms (they were costly to society), sometimes in social terms (they were a threat to the larger society), sometimes in terms of care (they needed to be protected *from* society). The upshot was the same. The people in the institutions were not valued, and others—voters, consumers—were. As James Trent notes in *Inventing the Feeble Mind*, the defunding of the institutions in New York, which led to the horrors of Willowbrook, followed the appropriation of funds for the Albany Mall.

Some institutions were abysmal, some mediocre, and some humane. They were called *training centers*, *schools*, *farms*, and *homes*, but "home" implies degrees of freedom the residents did not have. The institutions were less like homes than desert islands. In 1902, Trent reports, Martin Barr, superintendent of Pennsylvania's Elwyn Institute, proposed a literal island for the feeble-minded: "An ideal spot might be found—either on one of the newly acquired islands, the unoccupied lands of the Eastern seaboard, or the far West...." Anne Kerr and Tom Shakespeare report that in Denmark, in 1910, an actual island was procured for men with intellectual disabilities; a separate island for women was added in 1920. An island is the logical conclusion of the institution, a total separation. It is like a version of *The Tempest*, a place ruled by a powerful magician and separate from the ordinary flow of time, but built entirely around Caliban, around an idea of what is abnormal or monstrous.

I do not want Laura to live on an island, not even a pleasant one. I want the regular world to be good enough. But the appeal of a sanctuary, a place where she can live happily and the terrors of this world are barred, is impossible to deny. Even after World War II, when the sentiment toward "the retarded" began to turn, institutions continued to be built. Though the parents' movement of the 1950s marked a shift in the common sense of intellectual disability, many parents were focused on improving institutional conditions and parents' rights, not on civil rights for people with disabilities. The common sense I live in now, that I woke to when Laura arrived, is something I inherit from activists who came later, from self-advocates, and from parents who pushed for more than a separate place.

This is the backdrop against which I tell Laura's story. It is not in the past, and it is not ruled out from the future. To tell a different story is to oppose the assumptions in the dominant story and the injustices done in its name. Beyond that, I have more speculations than certainties, more questions than answers. A book is just a space where you can think about things a little differently, an unfamiliar house to walk through, a virtual home filled with the possibility of home.

We do not think that absolute security is possible or desirable. We do not want to secede or for Laura to be anywhere else. We have been lucky and hope our luck holds. We teach Laura to cross the street on her own. We let Laura cross the street on her own.

*   *   *

Intellect, normality, family, cost: these join the eugenic era to the present, like long continuous skeins, and persuasion weaves through them all like a bright ribbon through a braid. It links the state-run eugenics of the early twentieth century to the commercialized biomedicine of the twenty-first, the displays at Fitter Families Contests to today's ads for NIPT, and whatever CRISPR becomes or is succeeded by, persuasion will accompany that too. It's not inevitable that humans will be genetically engineered, but it is inevitable, given the level of investment in CRISPR, that it will become commercialized. For this reason, understanding the past of persuasion is useful for thinking about its future.

Eugenics and modern persuasion emerged concurrently, and they have a common root in industrial capitalism. The same force that sorted people in terms of economic viability, that made *ability* a matter of *ability to work*, that sped up the world and shone a spotlight on those judged slower, also drove the birth of advertising: for the economy to function, consumers needed to want the new things being produced. Their demand had to be aroused, channeled, and shaped. Despite the radical changes between then and now, much remains the same. The world is still swifter, still accelerating, and in a time of hyperidealized normality and achievement, people with intellectual disabilities are still left out and left behind. As for persuasion, the sophistication and reach of targeted online messaging exceeds anything the first advertisers could have possibly imagined.

With each technological innovation, Tim Wu writes in *The Attention Merchants*, the "industrial capture of attention" encroached further on our lives. Posters invaded public space. Radio crossed the threshold of the home: when radios first began to air advertisements, the home was thought to be sacrosanct, a space free from commercial intrusion. That belief quickly faded. TV arrived, followed by the personal computer, and now we have smartphones—"technological prosthetics," according to Wu, "enhancements of our own capacities, which, by virtue of being constantly attached to us or present on our bodies, become a part of us." Wu continues:

> And so, in the coming decade, the attention merchants will need to tread very lightly as they come as close as one can to the human body. Nonetheless, adaptation is a remarkable thing, and if our history has shown anything, it is that what seems shocking to one generation is soon taken for granted by the next.

I wonder if Wu's warning comes too late. I think of an image from Ariosa's website, part of the marketing for their Harmony prenatal test. A woman is in her home, her laptop open, holding a brochure. That is, a website meant to be viewed by women at their computers shows a picture of a woman at a computer. The brochure is significant: it shows that the model has *already* requested information, which is what the website hopes to convince the consumer to do. The model is one step into the future, a step ahead of the prospective consumer, a step closer to taking the test.

The kitchen the model sits in is palatial, filled with muted light. It is part of an implied narrative, a happy story built up in fragments: as you navigate from page to page, you see the same model held by an attractive man, or looking confidently into the camera, or filled with serenity, late in her pregnancy. This much echoes a long history of advertising, where the lives of beautiful, happy, well-off women are joined to a particular product—soap, cigarettes, mouthwash, antidepressants. What is different is that Ariosa's ad is also an implied narrative about information. Requesting information and receiving it are a prelude to taking the test, which means both giving information (personal data, DNA) and getting information (about the probable genetic status of a fetus). Behind this process is a series of transformations which depend, no less than the marketing, on computers: cell-free DNA is separated, translated into digital form, processed by proprietary algorithm, retranslated into a written report, and e-mailed to a doctor. Given the rapid transformations implied, the frictionless movement from sales pitch to blood draw, you could argue that in this case, the persuasion is already becoming part of us, that the attention merchants, having moved into our homes, have also crossed the boundary of our bodies.

Persuasion and eugenics emerged together, but until recently the two proceeded on separate tracks. Though eugenics faded, its central impulses did not disappear. As Nathaniel Comfort argues in *The Science of Human Perfection*, those impulses—health and human improvement—did not go away: they were dissolved into medical genetics. But with the rise of powerful computing, medical genetics and persuasion have converged in the realm of information. Genes, the technologies that "read" and "write" them, and the persuasion attached to the technologies, have a common home in computers.

Every technology transforms our understanding of ourselves and the world. Past technologies altered our understanding of time and distance, but if we take the next step beyond NIPT and embark on human gene editing, more than time and distance will be changed: when we *are* the project, we will become coextensive with the technologies we contemplate. Our code will live on computers and be rewritten there, and the rewritten code will live in us. Prenatal diagnosis changes the meaning of parenting; de-extinction changes the meaning of nature. But editing the germline would alter what it means to have a body, what it means for a body to be in a place. The meaning of bodies and homes.

And yet when I think about Laura's future—twenty years down the road, say—I worry less about a Gattaca world, and more about the slashing of the social safety net on which Laura will depend, if and when Theresa and I aren't around; I worry about what people will think about disability in general, whether the careful biosphere of welcome we've constructed around her will strengthen or collapse. It seems to me that the thinnest of transparent membranes shields her from the pure malice of this historical moment. She is happy, and she has the chance of happiness. But that is because we have the means to advocate for her, because a system exists within which she can have a good life, because we have found caring, amazing people in that system—teachers, administrators, physicians, specialists—who negotiate the realities of her condition to treat her as a person, rather than using the fact of the condition to dismiss her. If they do so, it's because they perceive her as more than an instance of misbegotten code, but as the possessor of a story.

* * *

I need a theory of relativity for understanding Down syndrome, the way time flows and eddies, uncertain. The way the observer's position alters the result. The kids Laura came up with in elementary school are driving and dating. Some friendships have faded, some have changed, and new ones have started. She and Maya sit in her room and talk and listen to Abba and color mandalas. She and Nina chatter at me over snacks, and Nina calls me Curious George, as she has for years, enormously pleased with herself as always. There are many ways of being in the world, and Laura's is a good one. I know she will have a place, but I don't know what it will look like. We are, as we always have been, on the edge of something new.

On a Tuesday afternoon I stand by the usual oak tree and wait for her to emerge from the school. I see her coming down the sidewalk, toting her heavy backpack, and she sees me and smiles in the specific shy way she has when I pick her up. She's happy to see me, but she's self-conscious in public. When she gets to the crosswalk, she waits carefully until a car comes to a full stop and a driver waves her across. We walk back to the car. I ask her if she has Spanish homework, and she says yes; I ask her if she wants help, and she says no; I say something silly, and she says No more sillies. Driving home, the phone peals twice and she picks it up and enters my passcode and taps on the message icon and reads a message from Theresa aloud. I say what to send back, and Laura pecks it out and presses send and clicks the screen dark. Simple, everyday gestures, from a future that's already here.

Acknowledgments

Thanks to those who read this book in manuscript, in whole or in part, and offered invaluable commentary: Rachel Adams, Emily Smith Beitiks, Shari Clough, Marcy Darnovsky, Theresa Filtz, Laura Mauldin, Justin St. Germain, Alexandra Minna Stern, Katie Stoll, and Megan Ward. In ways less direct but no less vital, I owe thanks to many for friendship, support, and conversations online and off that helped me develop these ideas: Anita Guerrini, Mike Osborne, Jim Trent, Tracy Daugherty, Allison Hobgood, Ed Hardy, Nicki Pombier-Berger, Jonathan Kaplan, Milton Reynolds, Beth Daley, Tim Jensen, Rosemarie Garland-Thomson, and Nathaniel Comfort. And to Alison Piepmeier, greatly missed.

Thanks to the editors of publications where material from this book first appeared, often in very different form: *Biopolitical Times, The Intima, Salon, Tin House, The AMA Journal of Ethics, The Oregonian,* and the *New York Times.* A special thanks to Peter Catapano at the *New York Times* for including my writing about Laura in the Disability series, and to Marcy Darnovsky for welcoming my writing at *Biopolitical Times.* Thanks also to Pete Shanks, Jessica Cussins, and everyone at the Center for Genetics and Society, present and past. And thanks to the National Society of Genetic Counselors for a grant to cover travel costs to their Annual Education Conference.

Lines from "Asphodel, That Greeny Flower," by William Carlos Williams, are reprinted by permission of New Directions Publishing Corp. from *The Collected Poems: Volume II, 1939–1962,* ©1944 by William Carlos Williams. Thanks to New Directions for permission to reprint. For permission to reprint these lines in the UK, many thanks to Carcanet Press Limited.

At the MIT Press, thanks to Bob Prior for rescuing this project from oblivion and for his incisive comments on the evolving draft. Thanks also to

Anne-Marie Bono for her quick reply to all manner of questions, to Deborah Cantor-Adams for moving the book efficiently through production, and to copy editor Regina Gregory for her sharp eye for detail. The book exists and is better for all of your efforts. Any errors that remain are mine alone.

Endless thanks to my dedicated agent and friend, Colleen Mohyde, who believed in this book, helped shape the proposal, and stuck with it until the (happy) end.

And finally, to my wife, Theresa Filtz, and my daughters, Ellie and Laura: This book is rooted in an understanding (vital, evolving, lived daily, mostly undefined) about family's possibilities. I owe that understanding, and my present happiness, to you.

Notes

Page vii   "Technology is neither good nor bad": Kranzberg (1986).

Page vii   "A world where everyone is welcome": Saxton (2017).

Introduction

Page xiii   Significant stretches of sequence: Genes also code for functional RNA molecules. On the evolving definitions of "gene," see Hopkin (2009).

Page xvii   CRISPR-Cas9: CRISPR stands for "clustered regularly interspaced short palindromic repeats," Cas9 for "CRISPR-associated protein 9."

Page xvii   "Far more, they are moments of provisionality": Jasanoff, Hurlbut, and Saha (2015).

Chapter 1: Virtual Children

Page 2   sexual violence against people with intellectual disabilities: Wissink et al. (2015).

Page 2   *Remaking Eden*: Silver (1997).

Page 4   the extraordinary, fine-grained act of control: For more on genes and parenting, see McKibben (2004).

Page 4   a beautiful, healthy baby and a laptop: GenePeeks (2012).

Page 4   *Science* profiled the company: Couzin-Frankel (2012).

Page 5   Andrew Solomon, writing in *The Guardian:* Solomon (2016).

Page 5   The feature article in *Science:* Couzin-Frankel (2012).

Page 6   Speaking to *New Scientist*: de Lange (2014).

Page 6   *Method and system for generating a virtual progeny genome*: Silver (2013).

Page 7   In her book *Feminist, Queer, Crip*: Kafer (2013).

Page 8   a YouTube video embedded: GenePeeks (2018).

Page 10   "One of the things I say a lot when I lecture": Hendren, quoted in Alvarez (2012).

Page 10   "After my son was born": Hendren, quoted in Collins (2017).

Page 11   "I want a technology that, yes, preserves independence": ibid.

Chapter 2: The Germline

Page 13   "Popular Opinion": Galton (1909).

Page 14   "the proportion of people of Grades A and B": Baynton (2011).

Page 14   Galton "first floated the idea for engineering society in 1865": Comfort (2014).

Page 16   Sometimes a cut occurs in the wrong place: Ravindran (2018).

Page 16   Two recent studies: Begley (2018).

Page 16   "standing on the cusp of a new era": Doudna and Sternberg (2017).

Page 16   the "dawn of the digital age": Venter (2014).

Page 16   "the greatest story ever told": Church and Regis (2014).

Page 16   revises religious language for secular ends: As Dorothy Nelkin pointed out in her essay "God Talk: Confusion between Science and Religion" (2004), sacred imagery informs the language of DNA, with the genome figured as Grail, Book of Life, the Code of Codes, and so on. That pattern is reflected in the titles of books about genes: *A Crack in Creation*, *Remaking Eden*, *Regenesis*. If the story's rhetoric leans toward the religious, it is likely because ultimate things are at stake.

Page 17   curable pathologies: Doudna and Sternberg (2017).

Page 18   using a patient's voice as a trump card: In Connor (2017), fertility researcher John Zhang, defending his use of an experimental IVF procedure performed in Mexico, said, "People who say we are working too fast must never have seen a family with a sick child in a wheelchair being fed through feeding tubes." For more on Zhang, see chapter 8.

Page 19   "Living at risk undermines confidence": Wexler (1996).

Page 19   everything from Tay-Sachs to deafness: Bérubé (2016a) notes the critical distinction between disability and disease, and the danger of conflating the two. "[E]ven in the most cure-averse precincts of disability studies," Bérubé writes, "there is no Polio Restoration Society, no Smallpox Appreciation League, no Cholera Pride movement." He notes that some conditions are difficult to categorize, but that "most disabilities have no disease etiologies whatsoever. Applying the cure/disease model to those disabilities is what philosophers call a category error, and fundamentally muddles our thinking about how to accommodate disability in society as best we can."

Page 19   In a famous TED talk: Young (2014).

Page 20   "The symptoms of hyperargininemia": Doudna and Sternberg (2017).

Page 20   In December of 2015: Weaver (2015).

Page 21   "Fertility companies freely admit": Cha (2017).

Page 22   "We may embrace much greater human diversity": Church and Regis (2014). Church's future scenario exemplifies what historian Allison Carey calls "controlled integration": "offering rights to people with disabilities based on their ability or potential to contribute" (Carey 2009). In this case, the right in question is the right to exist at all.

Page 23   "We call this collection of nearly all enzymes *E. pluri*": ibid.

Page 24   a sociotechnical system: Stevens (2016).

Page 24   *The Way Life Works*: Hoagland and Dodson (1995).

Page 25   the future holds "promise" and "peril": A Google search reveals a long list of titles, including "Gene Editing: Promise & Peril" (Meilaender 2017), "The Promise and Peril of CRISPR Gene Drives" (Zentner and Wade 2017), "CRISPR: The Promise and the Peril" (Scheufele 2016), "The Promise and Peril of Crispr" (Lauerman and Chen 2015), and so on.

Page 25   The fierce patent battles over CRISPR: Regalado (2014, 2017a).

Page 26   disability activist Rebecca Cokley: Cokley (2017).

Page 26   "People with disabilities are thinking about a traffic jam": Asch (1999).

Page 27   Most reject the simple equation of "abnormality" and "suffering": Garland-Thomson (2014).

Page 27   many reclaim a difference as a positive good, a different way of being in the world: As Laura Mauldin writes in *Made to Hear: Cochlear Implants and Raising Deaf Children,* to be Deaf is a cultural identity for many and not a disability: "Today there is extensive literature published on Deaf culture, a culture with its own history, language, and way of seeing the world.... Central to the Deaf critique is that they are a linguistic minority" (2016).

Page 27   as Rachel Kolb demonstrates: Kolb (2017).

Page 28   "the course of synthetic genomics": Church and Regis (2014).

Page 28   a man recovers spontaneously: Doudna and Sternberg (2017).

Page 29   David Starr Jordan ... "[an] aristocracy of brains": Solnit (2004). As with
            Galton, Jordan's eugenics were rooted in ideas about animal breeding
            and a hierarchical idea of human races. But as Alexandra Minna Stern
            writes (2015), Jordan's brand of eugenics was native to the American
            West. A love of the California landscape expressed the superiority of the
            whites who had occupied it but also fueled a hatred for those seen to
            threaten it, especially Mexicans and Asians. Eugenics was not a single
            thing: it evolved, taking on regional colorations.

Page 29   J. H. Kellogg: Selden (1999).

Page 29   "a special group of mental beings": Silver (1997).

Page 29   "Long after we are gone": Naam (2010).

Page 30   Galton's "galaxy of genius": Mark O'Connell (2017) notes the links
            between transhumanism, the dream of superior intellect, and space
            exploration.

Chapter 3: At the Fair

Page 31   "[T]he people of Oregon": Largent (2002).

Page 32   The directed evolution of people is inseparable from the directed evolution
            of animals: Bioethicist Paul Root Wolpe argues that we're entering a stage
            of directed evolution, and that our work on animals may pave the way
            for designer humans. As he notes, genetically modified food is ubiquitous,
            but further-out experiments have been accomplished: for example, neural
            implants in beetles (we can steer them in flight, using joysticks) or cats that
            glow under ultraviolet light (scientists have spliced genes from biolumines-
            cent jellyfish into Tabby's DNA, a literal Jellicle Cat). Wolpe (2010).

Page 32   CRISPR-engineered miniature pet pigs: Cyranoski (2015).

Page 33   "Purebred farm animals": Nelkin and Lindee (2004).

Page 33   A poster from the 1929 Kansas Free Fair: Selden (1999). Another display
            from the same fair: ibid.

Page 35   sterilizations ... in the California prison system: In 2010, the Center for
            Investigative Reporting exposed involuntary sterilizations at a California
            women's prison, performed without consent during routine operations.
            Dr. James Heinrich, the physician responsible for the sterilizations,

defended them explicitly in terms of cost, telling the Center for Investigative Reporting "that the money spent sterilizing inmates was minimal 'compared to what you save in welfare paying for these unwanted children—as they procreated more.'" Johnson (2013). In July of 2017, a Tennessee judge began giving inmates the option to reduce their sentences in exchange for permanent or long-term sterilization: men could get thirty days of freedom in exchange for a permanent vasectomy, and women were offered a long-term contraceptive implant. Hawkins (2017).

Page 35   "while the stock judges are testing the Holsteins": Boudreau (2005).

Page 36   a generic idea of the animal: In *White Trash: The Eugenic Family Studies 1877–1919,* Rafter (1988) writes that "[a]nimal and insect imagery pervades the family studies. The cacogenic 'mate' and 'migrate,' 'nesting' with their 'broods' in caves and 'hotbeds where human maggots are spawned.'…Most powerful of all is [the] extended metaphor of the 'pauper ganglion of several hundreds' that has attached itself to society as the Sacculina parasite attaches itself to the hermit crab, 'suck[ing] the living tissues.'"

Page 36   "animalization": O'Brien (2016).

Page 36   Leon Whitney: ibid.

Page 36   Charles Davenport, head of the Eugenics Record Office: ibid.

Page 36   "In other works he contended": ibid.

Page 36   "implored Hoosiers to adhere to the state's marriage laws": Stern (2002).

Page 37   "the fitter family contests": Lovett (2007).

Page 38   The bronze medal: Selden (1999).

Page 39   In *Regenesis*: Church and Regis (2014).

Page 40   "to enhance biodiversity": Revive & Restore (2018a).

Page 41   the Gene Security Network: Natera (2012).

Page 42   as Stewart Brand noted in a 2014 interview: Brand et al. (2014).

Page 43   "Such animals can also serve as icons," "Conservationists are learning the benefits," "The return of the marvelous marsupial wolf called the thylacine": Brand (2013).

Page 43   the intractable political, economic, and ecological complications: O'Connor (2015).

Page 43   "I found the individuals working on de-extinction projects to be brilliant": ibid.

Page 44   "[The mammoths'] return to the north would bring back carbon-fixing grass": Brand (2013).

Page 45   "Take hold of handles": Marzio (1973).

Page 46   The park is the brainchild of Sergey Zimov: Andersen (2017).

Page 46   "I will argue that the present focus": Shapiro (2016).

Page 47   in her account of the TEDx conference: for more on this conference, see Revive & Restore (2018b).

Page 47   The scientist is levelheaded and rational, the media and public ignorant and excitable: Shapiro (2016). See also Brand et al. (2014): "But immediately you would see in these wonderful comment lines after every place online, where the trolls emerge and start fretting, and the hand-wringing would be around, let's see: what if you bring it back and it turns out to be insanely invasive? And, passenger pigeons, there used to be five billion and suddenly there's five billion birds crapping on everything, it'll be like kudzu. Well, actually it's not an invasive. They were in North America for 22 million years. The invasive in this story is us, and we're the ones who shot them all to death. If we've got an invasive to worry about it's the human one, which is fair. Nature will accommodate these birds coming back. Nature's not broken."

Page 47   as anthropologist Sophia Roosth suggests: Roosth writes that the project's rhetoric is characterized by "the extrapolative and conjectural thinking that animates science fiction" (Roosth 2017).

Page 47   "adventure tourists," "Siberian eden": Church and Regis (2014).

Page 47   "bullet trains on elevated tracks": Andersen (2017).

Page 47   a pristine landscape: Shapiro (2016).

Page 47   Roosth argues that it's central: Roosth (2017).

Page 48   Live births of children with Down syndrome: de Graaf, Buckley, and Skotko (2018).

Page 48   extinct and threatened creatures as shadows in a landscape: Revive & Restore (2018a).

Chapter 4: On Our Screens

Page 52   "This is just a safer, more precise test": Abraham (2009).

Page 52   Stylli was ousted as CEO: Pollack (2009).

Page 52   From the Securities and Exchange Commission (SEC) report: Securities and Exchange Commission (2011). See also Securities and Exchange Commission (2010).

Page 53   the *Today* show profiled an expectant couple: "New Test Reveals Gender of Their Baby Live on Air" (2013).

Page 53   a range of health difficulties: Meredith (2017).

Page 53   a greatly reduced risk: Hasle et al. (2016).

Page 53   Babies born with trisomy 13 or trisomy 18: Brewer et al. (2002).

Page 54   an argument for inevitability: See also Maranto (2016).

Page 54   The existence of cell-free DNA: Lo et al. (1997).

Page 57   "Counter to their expectations, these mothers did not get what they worked for": Landsman (2008).

Page 57   In his classic history *Telethons:* Longmore (2016).

Page 58   In her memoir *Poster Child*: Rapp (2007).

Page 59   disability is more complex than supposed: In an essay written for *Slate* (2012), and in her second memoir, *At the Still Point of the Turning World,* Rapp (2013) narrates a very different story: the life and death of her son Ronan, who was born with Tay-Sachs disease, a fatal and progressive genetic illness. Most children who inherit the disease die before the age of three, as Ronan did, and though Rapp underwent prenatal testing, her son's rare variant was missed. In the *Slate* essay, Rapp writes beautifully about everything her son has brought to her life; she also writes that had his condition been detected, she would have aborted. She loves her son but does not see Tay-Sachs as a disability whose meaning depends on social acceptance; for her, it is a disease, and its primary meaning is suffering and early death. For more on Rapp and the categories of disability and disease, see Bérubé (2016a).

Page 59   "Waiter Hailed as Hero": Flam (2013).

Page 61   As Gail Landsman notes: Landsman (2008).

Page 63   Rahm Emanuel to Ann Coulter: Dwyer (2010); Grinberg (2012).

Page 63   "Know that the children of the world": Cho (2012).

Page 64   "I fear my body will have the last word": Cho (2012).

Page 64   "I think life is hard": Cho (2012).

Page 64   personal decisions about reproduction: Rayna Rapp (1999) writes
           that "[n]o one enters the decision to undergo amniocentesis trivially;
           genetic counseling is too sobering an experience to permit a casual
           use of this technology by any of the women among whom I have
           worked. In addition to its discursively enunciated risks, the counseling
           process is shadowed by the fears, fantasies, and phobias which preg-
           nant women and their supporters hold about childhood disabilities,
           and their pain at considering the possibility of ending a pregnancy to
           which they have already made a commitment.... The decision to use
           (or refuse) prenatal diagnosis engages deeply held values concerning
           the acceptable limits of maternity or parenthood, the importance of
           biomedical control over "nature," and the justification of abortion."

Page 65   I found a second clip from the after show: Cohen (2014).

Page 65   As I've written elsewhere: Estreich (2013). For a discussion of plagiarism
           in Down's ethnic theories, see Estreich (2014).

Page 67   "I'm a 30 year old man with Down syndrome": Stephens (2012).

Page 67   Mr. Stephens responded to comedian Gary Owen: Stephens (2016).

Page 67   Owen eventually met with a group of disability advocates: Diament (2016).

Chapter 5: The Fine Print

Page 69   an explanatory graphic: Natera (2014b).

Page 69   Sequenom's home page: Sequenom (2018b).

Page 70   the code we depend on is closed to us: O'Neil (2016).

Page 70   They look like models from a Cialis ad: Sequenom (2013).

Page 71   The image appears: DeBellis (2017), *Parents* magazine (2013), Morgan
           (2015).

Page 71   a picture of a child in a meadow: Natera (2014b), SafBaby (2013).

Page 72   Some Days I Feel Joyful: Progenity (2014); I Have Time to Lie Around
           in Meadows: Natera (2014a); I Have Money and I Like My Body: Ariosa
           Diagnostics (2017c); I've Never Experienced Nausea: Quest Diagnostics
           (2015); Namaste, NIPS!: Ariosa Diagnostics (2017a); A Male Model Loves
           and Protects Me: Ariosa Diagnostics (2017a); Verifi (2014).

Page 72   "As an expectant mother": Ariosa Diagnostics (2012a).

Page 73   "According to the American Congress of Obstetricians...": ibid.

Page 73   "Prenatal testing is part of almost every pregnancy": ibid.

Page 75   patent medicines were key to the rise of commercial persuasion: Wu (2017) writes, "It was…through the sale of patent medicine that advertising first proved conclusively its real utility, as a kind of alchemy, an apparently magical means of transforming basically useless substances into commercial gold."

Page 75   The ads for Listerine: ibid.

Page 75   the pioneering ad writer Helen Lansdowne: ibid.

Page 76   "As consumerism grew": ibid.

Page 77   "Down syndrome is a genetic disorder": Sequenom (2018a). "Women over the age of 35": ibid.

Page 77   In *Telling Genes*: Stern (2012).

Page 78   "the subtly misleading implications": Resta (2014).

Page 78   A web animation at natera.com: Natera (2018).

Page 79   Illumina, owner of the verifi test: Illumina (2017).

Page 79   *Find Clarity Early*: Ariosa Diagnostics (2017b). *Demand Clarity*: Ariosa Diagnostics (2017c). *Highly Accurate, Comprehensive Results You Can Trust*: Natera (2018). *The Reassurance of Knowing*: Verifi (2014). *Empowering Informed Choices*: Illumina (2017). *Pioneering Science, Personalized Service*: Sequenom (2018b). *Highly Accurate Answers to Important Questions*: Sequenom (2017).

Page 79   *Quality of Science*: Sequenom (2014a).

Page 80   Ariosa's explanation of the *Harmony* test: Ariosa Diagnostics (2017a).

Page 80   the phrase *your baby's DNA* implies the technology's reliability: that idea is reinforced graphically, with maternal and "fetal" DNA rendered in two different colors.

Page 81   a phenomenon called "confined placental mosaicism": See Lutgendorf et al. (2014).

Page 81   "a three-month examination": Daley (2014).

Page 82   Big Genoma: Resta (2018).

Page 83   *Better Results. Born of Better Science*: Sequenom (2014b).

Chapter 6: New Orleans

Page 86   "All the genetic counselors are women": Piepmeier (2013a).

Page 89   Increasingly, they're likely to work for one of the big labs: Resta (2018).

Page 90   the Natera-sponsored 22q11 Fun Run: Events like this are a part of a charity tradition. As Paul Longmore writes in *Telethons*, "It seemed that every health and disability charity in late-twentieth-century America put on vigorously physical fundraising events. There were walkathons and aerobathons and hopathons and bikeathons and danceathons for the March of Dimes, the UCP [United Cerebral Palsy], the American Cancer Society, the National Multiple Sclerosis Society, AIDS groups, the Cystic Fibrosis Foundation, the Leukemia Society, the Heart Fund, and more" (Longmore 2016).

Page 91   the cost of being a Silver Sponsor: National Society of Genetic Counselors (2015).

Page 91   a tried and true strategy of drug companies: "When pharmaceutical marketing does not look like science, it often looks like patient support" (Elliott 2003).

Page 91   as Katie Stoll pointed out: Stoll (2014).

Page 93   "whole-exome sequencing": National Institutes of Health (2018).

Page 94   faced with the overload of information: Condit (1999) writes, "[i]f commercial entities achieve their dream of multiplex testing, where prenatal scans include dozens or hundreds of potential genes, there will obviously be no way that the lay person can be said to have made an informed decision to choose those tests. Rational genetic choice for the individual will therefore become a practical impossibility."

Chapter 7: Reading Synthia

Page 95   "We stored information, and now we retrieved it": Regalado (2017b).

Page 95   Romesberg used a metaphor: Pollack (2014).

Page 95   In his TED Talk: Romesberg (2016).

Page 96   Romesberg's rhetoric walks a familiar line between old and new: Sophia Roosth writes that the rhetoric of synthetic biology "[wavers] between novelty and conventionality." In *Synthetic*, Roosth quotes an unnamed corporate executive who calls synthetic biology "schizophrenic": "... when we're talking to funders, we say synthetic biology is fundamentally different from biotech because we're making new things never found before in nature. But when we're talking to others, like the press, we say everything we're doing is *natural* and boring. We're just using natural things like plants, sugar, and yeast to make something useful" (Roosth 2017).

Page 96   "[h]istorians have documented": Stevens (2016).

Page 97   "There is no way of avoiding metaphors": Kay (1998).

Page 97   "The old metaphor is not wrong; it is incomplete": Comfort (2015).

Page 97   In a short article on genes and metaphor: Avise (2001).

Page 97   We do not know the function of every gene: Callaway (2016).

Page 98   "[I]f a genome is text": Comfort (2015).

Page 98   "the complexities of DNA's context-dependence": Kay (1998).

Page 98   "a message with a clear meaning, reliably reproduced": O'Keefe et al.
             (2015) argue that the "editing" metaphor is problematic: "'Editing' does
             not convey a sense of risk or a need for caution. It implies a 'mere text'
             that has an overall vision and a purpose within a bounded set of rules.
             Editors refine, correct, suggest, but they do so to improve. A text can be
             seen clearly; when a semicolon is changed to a colon, the grammatical
             function and the effect on meaning are known. But none of this is true
             of 'editing' a genome."

Page 98   In his book: Venter (2014). Synthia, announced in 2010, is part of an
             even more ambitious project: to build an organism with a "minimal
             gene set." In 2016, Venter announced the creation of Synthia 3.0. Ewen
             Callaway, writing in *Nature News* (2016): "Genomics entrepreneur Craig
             Venter has created a synthetic cell that contains the smallest genome
             of any known, independent organism. Functioning with 473 genes, the
             cell is a milestone in his team's 20-year quest to reduce life to its bare
             essentials and, by extension, to design life from scratch." According
             to the paper in *Science* detailing its creation, "JCVI-syn3.0 is a versatile
             platform for investigating the core functions of life and for explor-
             ing whole-genome design" (Hutchison et al. 2016). As the researchers
             also reported, the function of about a third of the organism's genes was
             unknown.

Page 99   As the *Guardian* reported: Sample (2010).

Page 99   as the *New York Times* reported: Wade (2010).

Page 100   Venter is relatively restrained: Gibson et al. (2010).

Page 100   Venter treats metaphor like an engineer stress-testing a metal: Venter
             (2014).

Page 101   "the proof…that DNA is the software of life": In *The Meanings of the
             Gene*, Celeste Condit (1999) comments on the code metaphor for DNA:
             "DNA, as bearer of the code, was not a mere messenger but a 'director'

or commander of the cell at large. Most significant, because it carried a code, DNA carried the *meaning* of the organism, and hence it provided the 'secret of life.'"

Page 102   the name was coined back in 2007: ETC Group (2007).

Page 102   named after Dolly Parton: Roslin Institute (2018).

Page 103   clearly implying that Venter belongs in their company: Daniel Falkner, commenting on Venter's 2012 address in Dublin called "What Is Life? A 21st-Century Perspective," notes that Venter's title echoed another lecture by Edwin Schrodinger—given "[a]t the same location, seventy years earlier," and also entitled "What Is Life?" As Falkner writes, Venter thereby "place[s] himself in an ancestral story of discovering the genetic code."

Page 103   using appropriated text to establish intellectual property rights: After Synthia was created, the Joyce estate sent Venter a cease-and-desist letter for quoting from Joyce's novel without authorization. Venter discussed this, and the "genetic typo," at SXSW. Ewalt (2011).

Page 104   "Synthetic biology is mostly about developing": Church and Regis (2014).

Page 104   "Richard Feynman issued a famous warning": Venter (2014). Venter's approving citation of this riff introduces a long passage in which he takes on his critics—specifically, those who say that Synthia is not a synthetic cell because Venter did not synthesize the entire cell. Venter acknowledges that definitions "are important in science," but also says that definitions "can be distractions that hinder how you think and what you do." He then argues that, by his definition, his cell qualifies as synthesized.

Page 105   simply lost: See Lepore (2015).

Page 106   "a borderless world": Venter (2014).

Page 106   *The future is already here*: Maharajh (2016).

Page 106   "the first self-replicating species": Wade (2010).

Page 109   "It was more of a symbolic idea of the world of science": Arad and Tolmach (2012).

Chapter 8: Dismissive Narratives

Page 113   "From thalidomide to global warming": Williams (2015).

Page 113   Harvard psychologist Steven Pinker: Pinker (2015).

Page 114   "life is better than death, health is better than disease, and vigor is better than disability": Knoepfler (2015b). Pinker's quote resembles one from Francis Galton: "All creatures would agree that it was better to be healthy than sick, vigorous than weak, well-fitted than ill-fitted for their part in life.... So with men" (Jones 2017).

Page 116   Over forty countries ban germline modification: Center for Genetics and Society (2015).

Page 116   in a common explanatory metaphor: Joseph (2016).

Page 117   mitochondria have complex regulatory functions: Hamilton (2014).

Page 117   "The Brave New World of Three-Parent I.V.F": Tingley (2014).

Page 118   a pattern familiar from websites advertising NIPT: In another parallel, problems associated with cytoplasmic transfer are acknowledged but de-emphasized. Tingley mentions briefly that two of the embryos created with cytoplasmic transfer had Turner syndrome, a chromosomal disorder in which an X chromosome is missing; one embryo miscarried, and the other was selectively terminated. (It's not known whether the disorder resulted from the procedure.)

Page 118   Philip Ball dismisses opponents: Ball (2014).

Page 119   At *Slate*: Grose (2014).

Page 119   In *The Guardian*: Toynbee (2014).

Page 119   "By constructing the public as ignorant": Wynne (2001).

Page 120   "When larger questions arise": Jasanoff, Hurlbut, and Saha (2015).

Page 120   "longstanding questions": Hurlbut (2015).

Page 121   an open letter to the parliamentary committee in the UK: Knoepfler (2014).

Page 122   an article in *Cell Stem Cell*: Yamada et al. (2016).

Page 122   "not well tolerated by normally fertilized zygotes": Hyslop et al. (2016).

Page 122   as Jessica Cussins noted: Cussins (2016).

Page 122   "clearly indicate that the field is not ready": Knoepfler (2016).

Page 123   Shoukhrat Mitalipov... published a paper in *Nature*: Kang et al. (2016).

Page 123   as Karen Weintraub reported in *Scientific American*: Weintraub (2016).

Page 123   the Human Fertilisation and Embryology Authority gave final approval: Human Fertilisation and Embryology Authority (2016).

Page 123   had already been born in Mexico: Hamzelou (2016).

Page 123   his cells showed evidence…his parents have resisted permission: Connor
            (2017).

Page 123   Around the same time, news arrived: Coghlan (2016).

Page 123   On their website: New Hope Fertility Center (2018).

Page 124   How will market pressures affect the science that supports the procedure:
            The authors of *CRISPR Democracy* comment on the patent battles over
            CRISPR, but what they say applies equally to three-person IVF: "With
            such forces in play, 'pushing research to its limits' easily translates into
            pushing biomedicine's commercial potential to its limits, meaning, in
            practice, that urgent needs of poor patients and overall public health
            may get sidelined in favor of developing non-essential treatments for
            affluent patients. Under these circumstances, it is hard not to read defenses
            of scientific autonomy and academic freedom as defenses of the freedom
            of the marketplace" (Jasanoff, Hurlbut, and Saha 2015).

Page 124   Stem-cell biologist George Daley: Reardon (2016b).

Page 124   Shoukhrat Mitalipov: Reardon (2016a). Dieter Egli, speaking to *Nature:*
            ibid.

Page 124   "I think the scientists and doctors showed a lot of courage": Reardon
            (2016b).

Page 125   "We'd love to do it with partners around the world": Reardon (2016a).

Page 126   her own person, a gift: See Sandel (2007).

Page 127   Taking folic acid is closer to the mark: genetic counselor Robert Resta
            (2002) describes "the use of folic acid to prevent neural tube defects,"
            along with initiatives like rubella vaccinations or patient education, as
            "non-eugenic": "counselors are trying to prevent the defect rather than
            the person with the defect."

Chapter 9: Model Worlds

Page 129   "The Search for Marvin Gardens": McPhee (1975).

Page 130   In 2013: Jiang et al. (2013).

Page 131   as science writer Ed Yong explains: Yong (2013).

Page 131   "Our hope," "Down's syndrome is a common disorder": Jiang et al.
            (2013).

Page 132   prenatal testing was justified: Stern (2012).

Page 132   the iconic number *35:* Resta (2002).

Page 133   a piece Alison wrote: Piepmeier (2013b).

Page 134   The "object lesson": Trent (2016).

Page 135   Philip Ferguson cites a report: Ferguson, Ferguson, and Brodsky (2008).

Page 135   "From the beginning": ibid.

Page 135   One Oregon family, writes Ferguson: ibid.

Page 136   In *The Gene: An Intimate History*: Mukherjee (2016). On stereotypes: Mukherjee also, in a doubtful metaphorical flourish, compares insulin (51 amino acids long) to somatostatin, "its duller, shorter cousin."

Page 136   the gulf in life expectancy: Santoro et al. (2016).

Page 137   The same pattern is evident in Andrew Solomon's *Far from the Tree:* Solomon (2012).

Page 137   One study notes relatively *low* rates of true aggression: Dykens (2007).

Page 138   "the danger of a single story": Adichie (2009).

Page 138   The release of a full report: Vargas (2013).

Page 138   His death was ruled a homicide: Rosenwald and Vargas (2013).

Page 138   Mr. Saylor's family has settled for $1.9 million: Vargas (2018).

Page 139   *The Selected Essays*: Cooley (2012).

Page 141   The sheriff, reacting to news of the independent investigation: Rosenwald and Vargas (2013).

Page 142   "It is difficult/to get the news from poems": Williams (1944).

Page 142   "Having special needs: it feels different": Cooley (2012).

Chapter 10: Finding a Place

Page 144   IDEA: Individuals with Disabilities Education Act (1975).

Page 145   "bearer of rights": Carey (2009).

Page 145   Douglas Baynton proposes a surprising argument: Baynton (2011).

Page 146   "Normal" is a particularly complex idea, embodying both "average and ideal": In *The Taming of Chance,* Ian Hacking writes, "One can, then, use the word 'normal' to say how things are, but also to say how they ought to be. The magic of the word is that we can use it to do both things at once." In the same chapter, Hacking continues: "The normal stands indifferently for what is typical, the unenthusiastic objective average, but it also stands for what has been, good health, and for what shall be, our chosen destiny. That is why the benign and sterile-sounding word 'normal' has become one of the most powerful ideological tools of the twentieth century" (Hacking 1990).

Page 147   "In the early twentieth century": Carey (2009).

Page 147   "[d]efects of the body, intellect, and moral sense": Baynton (2011).

Page 147   "the feebleminded woman": Carey (2009).

Page 147   "the borders needed to be guarded": In *Eugenic Nation*, Alexandra Minna Stern (2015) writes that "the Border Patrol was officially established on May 28, 1924. Its creation was motivated by the same eugenic arguments that pegged the quotas of the Johnson-Reed Immigration Act to the 1890 census." See also Baynton (2016).

Page 147   James Trent writes: Trent (2016).

Page 147   "There is no 'typical' citizen": Carey (2009).

Page 147   "a cruel and always a problematic faith": Kevles (1985).

Page 148   in the beginning was *Angel Unaware*: Rogers (1953).

Page 148   Rogers helped launch the parents' movement: This was literally true; the back flap to *Angel Unaware* noted that Rogers would donate all royalties to the National Association for Retarded Children (ibid.).

Page 151   as her biographer notes: White (2006).

Page 152   James Watson and Francis Crick: Watson and Crick (1953).

Page 152   Jérôme Lejeune would be credited: This discovery is disputed by another scientist, Marthe Gautier, who claims that she was the first to see the extra chromosome and that Lejeune appropriated her work. Pain (2014).

Page 153   "I grew up prizing intellectual aptitude": Gabbard (2010).

Page 155   Most adults with intellectual and developmental disabilities: The Arc (2011).

Page 155   exploitation: See, for example, Barry (2014).

Page 155   abuse: Vogell (2015).

Page 155   *The World of Nigel Hunt*: Hunt (1967).

Page 157   though this, too, is changing: Glatter (2017).

Conclusion: Bodies and Homes

Page 160   Muybridge's movie of a galloping horse: Shipman et al. (2017).

Page 160   Interviewed about the project, Shipman said: Yong (2017).

Page 161   Rebecca Solnit contends: Solnit (2004).

Page 162   the panoramic image of living and extinct species: Revive & Restore (2018b).

Page 162   In *Feminist, Queer, Crip*, Alison Kafer notes: Kafer (2013).

Page 163   "To become a maker": Solnit (2013).

Page 164   Here is the reformer Dorothea Dix: Quoted in Carey (2009).

Page 165   Eunice Kennedy Shriver: Shriver (1962).

Page 165   Michael Dukakis: Farragher (2017).

Page 165   From an article on Washington State institutions: "Conditions at State Institutions Unacceptable" (2017).

Page 166   the defunding of the institutions: Trent (2016).

Page 166   the horrors of Willowbrook: ibid. In 1972, Geraldo Rivera exposed horrific conditions in two New York State institutions, Willowbrook and Letchworth Village. According to Rivera, "[v]irtually every patient in building Tau was undressed and there was shit everywhere.…The residents of Tau were young girls. Many of them had physical deformities; most were literally smeared with feces—their roommates', their own. They looked like children who had been out making mudpies."

Page 166   a literal island for the feeble-minded: ibid.

Page 166   in Denmark, in 1910: Kerr and Shakespeare (2002).

Page 167   The parents' movement of the 1950s: Carey (2009).

Page 168   "And so, in the coming decade": Wu (2017).

Page 168   A woman is in her home: Ariosa Diagnostics (2017b).

Page 168   held by an attractive man: ibid.

Page 168   looking confidently into the camera: Ariosa Diagnostics (2017c).

Page 169   As Nathaniel Comfort argues: Comfort (2014).

References

Abraham, Carolyn. 2009. "Simple Test, Complex Questions." *The Globe and Mail*, February 7. https://www.theglobeandmail.com/news/national/simple-test-complex-questions/article1148236.

Adichie, Chimamanda Ngozi. 2009. "The Danger of a Single Story" (transcript). TEDGlobal 2009, October 7. https://www.ted.com/talks/chimamanda_adichie_the_danger_of_a_single_story/transcript.

Alvarez, Ana. 2012. "Inside the Prosthetic Imaginary: An Interview with Sara Hendren." *Rhizome*, October 4. http://rhizome.org/editorial/2012/oct/04/inside-prosthetic-imaginary-interview-sara-hendren.

Andersen, Ross. 2017. "Welcome to Pleistocene Park." *Atlantic*, April. https://www.theatlantic.com/magazine/archive/2017/04/pleistocene-park/517779.

Arad, Avi, and Matt Tolmach. 2012. "Arad and Tolmach Explain Why We Didn't See Norman Osborn in 'The Amazing Spider-Man.'" *Geekscape* (blog), July 5. http://www.geekscape.net/arad-and-tolmach-explain-why-we-didnt-see-norman-osborn-in-the-amazing-spider-man.

Arc, The. 2011. "Still in the Shadows with Their Future Uncertain." June. https://www.thearc.org/document.doc?id=3672.

Ariosa Diagnostics. 2012a. "Information for Pregnant Women." September 4. https://web.archive.org/web/20120904232539/http://www.ariosadx.com:80/for-pregnant-women.

Ariosa Diagnostics. 2012b. Home page. September 14. https://web.archive.org/web/20120914142002/http://www.ariosadx.com:80/#.

Ariosa Diagnostics. 2017a. "Expecting Parents." July 7. https://web.archive.org/web/20170707023810/http://www.ariosadx.com/expecting-parents.

Ariosa Diagnostics. 2017b. "Taking the Harmony Prenatal Test." July 7. https://web.archive.org/web/20170707040147/http://www.ariosadx.com/expecting-parents/taking-test.

Ariosa Diagnostics. 2017c. Home page. August 10. https://web.archive.org/web/2017 0810211228/http://www.ariosadx.com.

Asch, Adrienne. 1999. "Prenatal Diagnosis and Selective Abortion: A Challenge to Practice and Policy." *American Journal of Public Health* 89 (11): 1649–1657.

Asch, Adrienne. 2003. "Disability Equality and Prenatal Testing: Contradictory or Compatible?" *Florida State University Law Review. Florida State University. College of Law* 30 (2): 315–342.

Avise, John C. 2001. "Evolving Genomic Metaphors: A New Look at the Language of DNA." *Science* 294 (5540): 86–87. doi: 10.1126/science.294.5540.86.

Ball, Philip. 2014. "Unnatural Reactions." *The Lancet* 383 (9933): 1964–1965. doi:10.1016/S0140-6736(14)60945-4.

Barry, Dan. 2014. "The 'Boys' in the Bunkhouse." *New York Times*, March 8, sec. U.S. https://www.nytimes.com/interactive/2014/03/09/us/the-boys-in-the-bunkhouse .html.

Baynton, Douglas C. 2011. "'These Pushful Days': Time and Disability in the Age of Eugenics." *Health and History* 13 (2): 43–64.

Baynton, Douglas C. 2016. *Defectives in the Land: Disability and Immigration in the Age of Eugenics*. Chicago: University of Chicago Press.

Begley, Sharon. 2018. "CRISPR-Edited Cells Linked to Cancer Risk in 2 Studies." *Scientific American,* June 12. https://www.scientificamerican.com/article/crispr-edited-cells -linked-to-cancer-risk-in-2-studies.

Bérubé, Michael. 2016a. *Life as Jamie Knows It: An Exceptional Child Grows Up*. Boston: Beacon Press.

Bérubé, Michael. 2016b. *The Secret Life of Stories: From Don Quixote to Harry Potter, How Understanding Intellectual Disability Transforms the Way We Read*. New York: NYU Press.

Boudreau, Erica Bicchieri. 2005. "'Yea, I Have a Goodly Heritage': Health versus Heredity in the Fitter Family Contests, 1920–1928." *Journal of Family History* 30 (4): 366–387. doi:10.1177/0363199005276359.

Brand, Stewart. 2013. "Opinion: The Case for Reviving Extinct Species." *National Geographic News*, March 12. https://news.nationalgeographic.com/news/2013/03/1303 11-deextinction-reviving-extinct-species-opinion-animals-science.

Brand, Stewart, Richard Prum, Hans Ulrich Obrist, and John Brockman. 2014. "De-Extinction." October 31. https://www.edge.org/conversation/de-extinction-stewart -brand-richard-prum-with-hans-ulrich-obrist-and-brockman-part-i.

Brewer, C. M., S. H. Holloway, D. H. Stone, A. D. Carothers, and D. R. Fitzpatrick. 2002. "Survival in Trisomy 13 and Trisomy 18 Cases Ascertained from Population

Based Registers." *Journal of Medical Genetics* 39 (9). https://jmg.bmj.com/content/39/9/e54.

Callaway, Ewen. 2016. "'Minimal' Cell Raises Stakes in Race to Harness Synthetic Life." *Nature News* 531 (7596): 557. doi: 10.1038/531557a.

Carey, Allison C. 2009. *On the Margins of Citizenship: Intellectual Disability and Civil Rights in Twentieth-Century America.* Philadelphia: Temple University Press.

Center for Genetics and Society. 2015. Human Germline Modification: Summary of National and International Policies. June. https://www.geneticsandsociety.org/sites/default/files/cgs_global_policies_summary_2015.pdf.

Cha, Ariana Eunjung. 2017. "Discounts, Guarantees and the Search for 'Good' Genes: The Booming Fertility Business." *The Washington Post*, October 21, sec. Health Science. https://www.washingtonpost.com/national/health-science/donor-eggs-sperm-banks-and-the-quest-for-good-genes/2017/10/21/64b9bdd0-aaa6-11e7-b3aa-c0e2e1d41e38_story.html.

Cho, Margaret. 2012. "More Apologies." *Margaret Cho Official Site* (blog), June 1. http://margaretcho.com/2012/06/01/more-apologies.

Church, George M., and Ed Regis. 2014. *Regenesis: How Synthetic Biology Will Reinvent Nature and Ourselves.* New York: Basic Books.

Coghlan, Andy. 2016. "'3-Parent' Baby Method Already Used for Infertility." *New Scientist,* October 10. https://www.newscientist.com/article/2108549-exclusive-3-parent-baby-method-already-used-for-infertility.

Cohen, Andy. 2014. "After Show: Rosie Pope on 'RHONJ' Parenting." *Watch What Happens Live.* http://www.bravotv.com/watch-what-happens-live/season-6/videos/after-show-rosie-pope-on-rhonj-parenting.

Cokley, Rebecca. 2017. "Please Don't Edit Me Out." *Washington Post*, August 10, sec. Opinions. https://www.washingtonpost.com/opinions/if-we-start-editing-genes-people-like-me-might-not-exist/2017/08/10/e9adf206-7d27-11e7-a669-b400c5c7e1cc_story.html?utm_term=.337a65a9cde8.

Collins, Gillie. 2017. "Sara Hendren: The Body Adaptive." *Guernica*, February 6. https://www.guernicamag.com/sara-hendren-the-body-adaptive.

Comfort, Nathaniel. 2014. *The Science of Human Perfection: How Genes Became the Heart of American Medicine.* Reprint edition. New Haven, CT: Yale University Press.

Comfort, Nathaniel. 2015. "Genetics: We Are the 98%." *Nature* 520 (7549): 615–616. doi:10.1038/520615a.

Condit, Celeste. 1999. *The Meanings of the Gene: Public Debates about Human Heredity.* Madison: University of Wisconsin Press.

"Conditions at State Institutions Unacceptable." 2017. Editorial Board, *The News-Tribune*, August 9. http://www.thenewstribune.com/opinion/article166391752.html.

Connor, Steve. 2017. "When Replacement Becomes Reversion." *Nature Biotechnology,* November 9. doi: 10.1038/nbt.3996.

Cooley, Sarah. 2012. *The Selected Essays of Sarah Savage Cooley.* 2nd ed. Edited by Reed Cooley and Lucy Flores. San Francisco: Blurb Inc.

Couzin-Frankel, Jennifer. 2012. "New Company Pushes the Envelope on Pre-conception Testing." *Science* 338 (6105): 315–316. doi:10.1126/science.338.6105.315.

Cussins, Jessica. 2016. "UK Researchers Now Say Three-Person Embryo Technique Doesn't Work; Propose New Method." Center for Genetics and Society. June 8. https://www.geneticsandsociety.org/biopolitical-times/uk-researchers-now-say-three-person-embryo-technique-doesnt-work-propose-new.

Cyranoski, David. 2015. "Gene-edited 'Micropigs' to Be Sold as Pets at Chinese Institute." *Nature News* 526 (7571): 18. doi: 10.1038/nature.2015.18448.

Daley, Beth. 2014. "Oversold Prenatal Tests Leading to Abortions." *Boston Globe*, December 14. https://www.bostonglobe.com/metro/2014/12/14/oversold-and-unregulated-flawed-prenatal-tests-leading-abortions-healthy-fetuses/aKFAOCP5N0Kr8S1HirL7EN/story.html.

DeBellis, Lauren. 2017. "A Guide to Your Preconception Visit." Parents. April 21. Accessed January 8, 2018. https://www.parents.com/getting-pregnant/pre-pregnancy-health/general/your-preconception-doctor-visit-guide.

de Graaf, Gert, Frank Buckley, and Brian Skotko. 2018. People Living with Down Syndrome in the USA: Births and Population. Fact sheet, updated February 21. https://assets.cdn.down-syndrome.org/files/reports/research/births-prevalence/usa/down-syndrome-population-usa-factsheet-20180221.pdf.

de Lange, Catherine. 2014. "Meet Your Unborn Child—Before It's Even Conceived." *New Scientist*, April 9. https://www.newscientist.com/article/mg22229642-800-meet-your-unborn-child-before-its-even-conceived.

Diament, Michelle. 2016. "'Retarded' Comedy Routine Pulled from Showtime." Disability Scoop, June 9. https://www.disabilityscoop.com/2016/06/09/retarded-pulled-showtime/22391.

Doudna, Jennifer A., and Samuel H. Sternberg. 2017. *A Crack in Creation: Gene Editing and the Unthinkable Power to Control Evolution.* Boston: Houghton Mifflin Harcourt.

Dwyer, Devin. 2010. "Rahm Emanuel Puts 'Retarded' Offensiveness in Spotlight." ABC News. February 4. https://abcnews.go.com/WN/rahm-emanuel-retarded-comment-puts-offensiveness-spotlight/story?id=9738134.

Dykens, Elisabeth M. 2007. "Psychiatric and Behavioral Disorders in Persons with Down Syndrome." *Mental Retardation & Developmental Disabilities Research Reviews* 13 (3): 272–278. doi: 10.1002/mrdd.20159.

Elliott, Carl. 2003. *Better than Well: American Medicine Meets the American Dream.* New York: W. W. Norton.

Estreich, George. 2013. *The Shape of the Eye: A Memoir.* New York: Tarcher/Penguin.

Estreich, George. 2014. "Richard Dawkins Gets It All Wrong, Yet Again." *Salon,* September 25. https://www.salon.com/2014/09/25/richard_dawkins_gets_it_all_wrong_yet_again.

ETC Group. 2007. "Patenting Pandora's Bug: Goodbye, Dolly… Hello, Synthia!" ETC Group. June 7. http://www.etcgroup.org/content/patenting-pandora%E2%80%99s-bug -goodbye-dollyhello-synthia.

Ewalt, David M. 2011. "Craig Venter's Genetic Typo." *Forbes.* March 14. https:// www.forbes.com/sites/davidewalt/2011/03/14/craig-venters-genetic-typo.

Falkner, Daniel. 2016. "Metaphors of Life: Reflections on Metaphors in the Debate on Synthetic Biology." In *Ambivalences of Creating Life: Societal and Philosophical Dimensions of Synthetic Biology,* ed. Kristin Hagen, Kristin, Margret Engelhard, and Georg Toepfer, 251–265. Basel: Springer International.

Farragher, Thomas. 2017. "Mike Dukakis: From Brink of the Presidency to a Quiet Life of Significance." *Boston Globe,* September 8. https://www.bostonglobe.com/metro /2017/09/08/mike-dukakis-from-brink-presidency-quiet-life-significance/wLvqfMLr VS31tHWmXfmKgJ/story.html.

Ferguson, Philip, Dianne L. Ferguson, and Meredith M. Brodsky. 2008. *"Away from the Public Gaze": A History of the Fairview Training Center and the Institutionalization of People with Developmental Disabilities in Oregon.* The Teaching Research Institute, Western Oregon University. https://digital.osl.state.or.us/islandora/object/osl%3A 22866/datastream/OBJ/view.

Flam, Lisa. 2013. "Waiter Hailed as Hero after Standing Up for Boy with Down Syndrome." Today.com, January 23. https://www.today.com/parents/waiter-hailed -hero-after-standing-boy-down-syndrome-1B8038223.

Gabbard, Chris. 2010. "A Life beyond Reason." *Chronicle of Higher Education,* November 7. https://www.chronicle.com/article/A-Life-Beyond-Reason/125242.

Galton, Francis. 1909. *Essays in Eugenics.* London: Eugenics Education Society.

Garland-Thomson, Rosemarie. 2014. "The Story of My Work: How I Became Disabled." *Disability Studies Quarterly* 34 (2). http://dsq-sds.org/article/view/4254.

GenePeeks. 2012. Home page, December 1. https://web.archive.org/web/201212011 00011/http://www.genepeeks.com:80.

GenePeeks. 2018. Home page, January 5. https://web.archive.org/web/20180105 203251/https://www.genepeeks.com.

Gibson, Daniel G., John I. Glass, Carole Lartigue, Vladimir N. Noskov, Ray-Yuan Chuang, Mikkel A. Algire, Gwynedd A. Benders, et al. 2010. "Creation of a Bacterial Cell Controlled by a Chemically Synthesized Genome." *Science* 329 (5987): 52–56. doi:10.1126/science.1190719.

Glatter, Hayley. 2017. "The Path to Higher Education with an Intellectual Disability." *Atlantic*, May 1. https://www.theatlantic.com/education/archive/2017/05/the-path -to-higher-education-with-an-intellectual-disability/524748.

Grinberg, Emanuella. 2012. "Ann Coulter's Backward Use of the 'r-Word.'" CNN. October 24. https://www.cnn.com/2012/10/23/living/ann-coulter-obama-tweet/index .html.

Grose, Jessica. 2014. "'Designer Babies' Aren't Coming. The New York Times Is Just Trying to Scare You." *Slate*, February 26. http://www.slate.com/blogs/xx_factor/2014 /02/26/_designer_babies_aren_t_on_their_way_hopefully_3_person_embryo_fertiliza tion.html.

Hacking, Ian. 1990. *The Taming of Chance.* Cambridge: Cambridge University Press.

Hamilton, Garry. 2014. "Possessed! The Powerful Aliens That Lurk within You." *New Scientist*, September 17. https://www.newscientist.com/article/mg22329870-600 -possessed-the-powerful-aliens-that-lurk-within-you.

Hamzelou, Jessica. 2016. "Exclusive: World's First Baby Born with New '3 Parent' Technique." *New Scientist*, September 27. https://www.newscientist.com/article/2107 219-exclusive-worlds-first-baby-born-with-new-3-parent-technique.

Hasle, Henrik, Jan M. Friedman, Jorgen H. Olsen, and Sonja Rasmussen. 2016. "Low Risk of Solid Tumors in Persons with Down Syndrome." *Genetics in Medicine* (March 31). https://doi-org.ezproxy.proxy.library.oregonstate.edu/10.1038/gim.2016.23.

Hawkins, Derek. 2017. "Judge to inmates: Get sterilized and I'll shave off jail time." *The Washington Post,* sec. Morning Mix, July 21. https://www.washingtonpost.com /news/morning-mix/wp/2017/07/21/judge-to-inmates-get-sterilized-and-ill-shave -off-jail-time.

Hoagland, Mahlon, and Bert Dodson. 1995. *The Way Life Works.* 1st ed. New York: Crown.

Hopkin, Karen. 2009. "The Evolving Definition of a Gene: With the Discovery That Nearly All of the Genome Is Transcribed, the Definition of a 'Gene' Needs Another Revision." *BioScience* 59 (11): 928–931. doi: 10.1525/bio.2009.59.11.3.

Human Fertilisation and Embryology Authority, Strategy and Information Director- ate. 2016. "HFEA Permits Cautious Use of Mitochondrial Donation in Treatment,

Following Advice from Scientific Experts." https://www.hfea.gov.uk/about-us/news
-and-press-releases/2016-news-and-press-releases/hfea-permits-cautious-use-of
-mitochondrial-donation-in-treatment-following-advice-from-scientific-experts.

Hunt, Nigel. 1967. *The World of Nigel Hunt: The Diary of a Mongoloid Youth.* New
York: Taplinger.

Hurlbut, J. Benjamin. 2015. "Limits of Responsibility: Genome Editing, Asilomar,
and the Politics of Deliberation." *Hastings Center Report* 45 (5): 11–14. doi:10.1002/
hast.484.

Hutchison, Clyde, Ray-Yuan Chuang, Vladimir N. Noskov, Nacyra Assad-Garcia,
et al. 2016. "Design and Synthesis of a Minimal Bacterial Genome." *Science* 351 (6280).
doi: 10.1126/science.aad6253.

Hyslop, Louise A., Paul Blakeley, Lyndsey Craven, Jessica Richardson, Norah M. E.
Fogarty, Elpida Fragouli, Mahdi Lamb, et al. 2016. "Towards Clinical Application of
Pronuclear Transfer to Prevent Mitochondrial DNA Disease." *Nature* 534 (7607): 383.
doi:10.1038/nature18303.

iBiology. 2015. "CRISPR: A Word Processor for Editing the Genome." IBiology.
March 2015. https://www.ibiology.org/genetics-and-gene-regulation/crispr.

Illumina. 2017. "Noninvasive Prenatal Testing (NIPT)." July 12. https://web.archive
.org/web/20170712114202/https://www.illumina.com/clinical/reproductive-genetic
-health/nipt.html.

Individuals with Disabilities Education Act. 1975. Accessed January 12, 2018. https://
sites.ed.gov/idea/about-idea/#IDEA-History.

Jasanoff, Sheila, J. Benjamin Hurlbut, and Krishanu Saha. 2015. "CRISPR Democracy: Gene Editing and the Need for Inclusive Deliberation." *Issues in Science and
Technology* 32 (1). http://issues.org/32-1/crispr-democracy-gene-editing-and-the-need
-for-inclusive-deliberation.

Jiang, Jun, Yuanchun Jing, Gregory J. Cost, Jen-Chieh Chiang, Heather J. Kolpa, Allison M. Cotton, Dawn M. Carone, et al. 2013. "Translating Dosage Compensation to
Trisomy 21." *Nature* 500 (7462): 296. doi:10.1038/nature12394.

Johnson, Corey G. 2013. "Female Inmates Sterilized in California Prisons without
Approval." *Reveal,* from the Center for Investigative Reporting. July 7. https://www
.revealnews.org/article/female-inmates-sterilized-in-california-prisons-without
-approval.

Jones, Sarah. 2017. "Trump Has Turned the GOP into the Party of Eugenics." *The
New Republic,* February 15. https://newrepublic.com/article/140641/trump-turned
-gop-party-eugenics.

Joseph, Andrew. 2016. "World's First Baby Born with Novel Three-Parent Embryo Technique." *Stat,* September 27. https://www.statnews.com/2016/09/27/three-parent -baby-embryo.

Kafer, Alison. 2013. *Feminist, Queer, Crip.* Bloomington: Indiana University Press.

Kang, Eunju, et al. 2016. "Mitochondrial Replacement in Human Oocytes Carrying Pathogenic Mitochondrial DNA Mutations." *Nature* 540 (7632): 270–275.

Kay, Lily E. 1998. "A Book of Life? How the Genome Became an Information System and DNA a Language." *Perspectives in Biology and Medicine* 41 (4): 504–528.

Kerr, Anne, and Tom Shakespeare. 2002. *Genetic Politics: From Eugenics to Genome.* Cheltenham, UK: New Clarion Press.

Kevles, Daniel J. 1985. *In the Name of Eugenics: Genetics and the Uses of Human Heredity.* 1st ed. New York: Knopf.

Knoepfler, Paul. 2014. "Open Letter to UK Parliament: Avoid Historic Mistake on Rushing Human Genetic Modification." *The Niche* (blog), November 2. https://ips cell.com/2014/11/open-letter-to-uk-parliament-avoid-historic-mistake-on-rushing -human-genetic-modification.

Knoepfler, Paul. 2015a. *GMO Sapiens: The Life-Changing Science of Designer Babies.* 1st ed. Hackensack, NJ: World Scientific.

Knoepfler, Paul. 2015b. "Steven Pinker Interview: Case against Bioethocrats & CRISPR Germline Ban." *The Niche* (blog), August 10. https://ipscell.com/2015/08 /stevenpinker.

Knoepfler, Paul. 2016. "New Herbert Lab Nature Paper Reinforces Mitochondrial Replacement Achilles Heel." *The Niche* (blog), June 8. https://ipscell.com/2016/06 /new-nature-paper-reinforces-that-mitochondrial-replacement-has-achilles-heel.

Kolb, Rachel. 2017. "Sensations of Sound: On Deafness and Music." *New York Times,* November 3, sec. Opinion. https://www.nytimes.com/2017/11/03/opinion/cochlear -implant-sound-music.html.

Kranzberg, Melvin. 1986. "Technology and History: 'Kranzberg's Laws.'" *Technology and Culture* 27 (3): 544–560. doi:10.2307/3105385.

Landsman, Gail. 2008. *Reconstructing Motherhood and Disability in the Age of Perfect Babies.* London: Routledge.

Largent, Mark. 2002. "'The Greatest Curse of the Race': Eugenic Sterilization in Oregon, 1909–1983." *Oregon Historical Quarterly* 103 (2) (Summer): 188–209.

Lauerman, John, and Caroline Chen. 2015. "The Promise and Peril of Crispr." *Bloomberg.Com,* June 25. https://www.bloomberg.com/news/articles/2015-06-25/the -promise-and-peril-of-crispr.

Ledford, Heidi. 2017. "CRISPR Fixes Disease Gene in Viable Human Embryos." *NAT-News* 548 (7665): 13. doi:10.1038/nature.2017.22382.

Lepore, Jill. 2015. "What the Web Said Yesterday." *The New Yorker*, January 19. https://www.newyorker.com/magazine/2015/01/26/cobweb.

Lo, Y. M. Dennis, Noemi Corbetta, Paul F. Chamberlain, Vik Rai, Ian L. Sargent, Christopher W.G. Redman, and James S. Wainscoat. 1997. "Presence of Fetal DNA in Maternal Plasma and Serum." *The Lancet* 350 (9076): 485.

Longmore, Paul K. 2016. *Telethons: Spectacle, Disability, and the Business of Charity.* Oxford: Oxford University Press.

Lovett, Laura L. 2007. "'Fitter Families for Future Firesides': Florence Sherbon and Popular Eugenics." *Public Historian* 29 (3): 69–85. doi:10.1525/tph.2007.29.3.69.

Lutgendorf, Monica A., Katie A. Stoll, Dana M. Knutzen, and Lisa M. Foglia. 2014. "Noninvasive Prenatal Testing: Limitations and Unanswered Questions." *Genetics in Medicine* 16 (4): 281–285. doi: 10.1038/gim.2013.126.

Maharajh, Robert. 2016. "The Future Has Arrived." *Not Evenly Distributed* (blog), May 24. https://medium.com/not-evenly-distributed/the-future-has-arrived-fed56cec3266.

Maranto, Gina. 2016. "False Inevitabilities and Irrational Exuberance." Biopolitical Times. January 8. https://www.geneticsandsociety.org/biopolitical-times/false-inevitabilities-and-irrational-exuberance.

Marzio, Peter C. 1973. *Rube Goldberg: His Life and Work.* 1st ed. New York: Harper & Row.

Mauldin, Laura. 2016. *Made to Hear: Cochlear Implants and Raising Deaf Children.* Minneapolis: University of Minnesota Press.

McKibben, Bill. 2004. *Enough: Staying Human in an Engineered Age.* New York: St. Martin's Griffin.

McPhee, John. 1975. "The Search for Marvin Gardens." In *Pieces of the Frame.* New York: Farrar, Straus, and Giroux.

Meilaender, Gilbert. 2017. "Gene Editing: Promise & Peril." March 13. https://www.commonwealmagazine.org/gene-editing-promise-peril.

Meredith, Stephanie. 2017. "Medical Issues." In *Understanding a Down Syndrome Diagnosis.* Joseph P. Kennedy, Jr. Foundation and the Human Development Institute. http://understandingdownsyndrome.org/slides/4.

Morgan, Mandy. 2015. "Parents Using Infertility Treatment to Determine Sex of Their Child." DeseretNews.com. August 25. https://www.deseretnews.com/article/865635281/Parents-who-have-no-problem-getting-pregnant-still-turn-to-fertility-treatments-for-help.html.

Mukherjee, Siddhartha. 2016. *The Gene: An Intimate History.* 1st ed. New York: Scribner.

Naam, Ramez. 2010. *More Than Human: Embracing the Promise of Biological Enhancement.* New York: lulu.com.

Natera. 2012. "PGD Testing, NIPD Testing, Miscarriage Testing." March 22. https:// web.archive.org/web/20120322205300/http://www.natera.com.

Natera. 2014a. "Providers." August 29. https://web.archive.org/web/20140829174847 /http://www.panoramatest.com/en/healthcare-provider.

Natera. 2014b. "Expecting Parents." December 8. https://web.archive.org/web /20141208161254/http://www.panoramatest.com:80/expecting-mother.

Natera. 2018. "What Is Panorama?" January 9. https://web.archive.org/web/2018010 9034818/https://www.natera.com/panorama-test.

National Institutes of Health. 2018. "What Are Whole Exome Sequencing and Whole Genome Sequencing?" Genetics Home Reference, July 17. https://ghr.nlm .nih.gov/primer/testing/sequencing.

National Society of Genetic Counselors. 2015. "Elevate Your Brand." https://www .nsgc.org/page/elevate-your-brand.

Nelkin, Dorothy. 2004. "God Talk: Confusion between Science and Religion." *Science, Technology, and Human Values*, April 1. doi: 10.1177/0162243903261950.

Nelkin, Dorothy, and M. Susan Lindee. 2004. *The DNA Mystique: The Gene as a Cultural Icon.* Ann Arbor: University of Michigan Press.

New Hope Fertility Center. 2018. "Best Fertility Clinic NYC—IVF Treatment Protocol— Hope Fertility Center." New Hope Fertility Center. Accessed January 11, 2018. https:// www.newhopefertility.com.

"New Test Reveals Gender of Their Baby Live on Air." 2013, January 28. https:// www.msnbc.com/today/watch/new-test-reveals-gender-of-their-baby-live-on-air -16498755506.

O'Brien, Gerald V. 2016. *Framing the Moron: The Social Construction of Feeble-Mindedness in the American Eugenic Era.* Reprint edition. Manchester: Manchester University Press.

O'Connell, Mark. 2017. *To Be a Machine: Adventures among Cyborgs, Utopians, Hackers, and the Futurists Solving the Modest Problem of Death.* New York: Doubleday.

O'Connor, M. R. 2015. *Resurrection Science: Conservation, De-Extinction and the Precarious Future of Wild Things.* New York: St. Martin's Press.

O'Keefe, Meaghan, Sarah Perrault, Jodi Halpern, Lisa Ikemoto, and Mark Yarborough. 2015. "'Editing' Genes: A Case Study about How Language Matters in Bioethics." *The American Journal of Bioethics,* 15 (12): 3–10. doi: 10.1080/15265161.2015.1103804.

O'Neil, Cathy. 2016. *Weapons of Math Destruction: How Big Data Increases Inequality and Threatens Democracy.* New York: Broadway Books.

Pain, Elisabeth. 2014. "After More Than 50 Years, a Dispute over Down Syndrome Discovery." *Science,* February 11. http://www.sciencemag.org/news/2014/02/after-more -50-years-dispute-over-down-syndrome-discovery.

Parents magazine. 2013. "Should You Be Evaluated for Infertility?" *Parents,* October 23. https://www.parents.com/getting-pregnant/infertility/treatments/infertility -evaluation.

Pernick, Martin S. 1996. *The Black Stork: Eugenics and the Death of "Defective" Babies in American Medicine and Motion Pictures since 1915.* 1st ed. New York: Oxford University Press.

Piepmeier, Alison. 2013a. "Every Little Thing: Feminist Disability Studies Scholar Attends the NSGC." *Every Little Thing* (blog), October 11. http://alisonpiepmeier .blogspot.com/2013/10/feminist-disability-studies-scholar.html.

Piepmeier, Alison. 2013b. "Outlawing Abortion Won't Help Children with Down Syndrome." *New York Times,* sec. Motherlode (blog), April 1. https://parenting.blogs .nytimes.com/2013/04/01/outlawing-abortion-wont-help-children-with-down -syndrome.

Pinker, Steven. 2015. "The Moral Imperative for Bioethics." *Boston Globe,* August 1. https://www.bostonglobe.com/opinion/2015/07/31/the-moral-imperative-for-bioethics /JmEkoyzlTAu9oQV76JrK9N/story.html.

Pollack, Andrew. 2009. "Sequenom Fires Chief and Others over Handling of Data." *New York Times,* September 28, sec. Business Day. https://www.nytimes.com/2009 /09/29/business/29drug.html.

Pollack, Andrew. 2014. "Scientists Add Letters to DNA's Alphabet, Raising Hope and Fear." *New York Times,* May 7, sec. Business Day. https://www.nytimes.com/2014 /05/08/business/researchers-report-breakthrough-in-creating-artificial-genetic-code .html.

Progenity. 2014. "Progenity." August 4. https://web.archive.org/web/20140804010915 /https://progenity.com.

Quest Diagnostics. 2015. "Prenatal: About Noninvasive Prenatal Screening." July 17. https://web.archive.org/web/20150717121655/http://www.questdiagnostics.com:80 /home/patients/tests-a-z/prenatal/during-pregnancy/noninvasive/about.

Rafter, Nicole. 1988. *White Trash: The Eugenic Family Studies 1877–1919*. Boston: Northeastern University Press.

Rapp, Emily. 2007. *Poster Child: a Memoir*. New York: Bloomsbury.

Rapp, Emily. 2012. "Rick Santorum, Meet My Son." *Slate,* February 27. http://www.slate.com/articles/double_x/doublex/2012/02/rick_santorum_and_prenatal _testing_i_would_have_saved_my_son_from_his_suffering_.html.

Rapp, Emily. 2013. *At the Still Point of the Turning World*. New York: Penguin.

Rapp, Rayna. 1999. *Testing Women, Testing the Fetus: The Social Impact of Amniocentesis in America*. New York: Routledge.

Ravindran, Sandeep. 2018. "New Methods to Detect CRISPR Off-Target Mutations." *The Scientist,* March 1. https://www.the-scientist.com/lab-tools/new-methods-to-detect -crispr-off-target-mutations-30013.

Reardon, Sara. 2016a. "'Three-Parent Baby' Claim Raises Hopes—and Ethical Concerns." *Nature,* September 28. doi:10.1038/nature.2016.20698.

Reardon, Sara. 2016b. Reports of 'Three-Parent Babies' Multiply. *Nature,* October 19. doi:10.1038/nature.2016.20849.

Regalado, Antonio. 2014. "Who Owns the Biggest Biotech Discovery of the Century?" MIT Technology Review. December 4. https://www.technologyreview.com/s /532796/who-owns-the-biggest-biotech-discovery-of-the-century.

Regalado, Antonio. 2017a. "Broad Institute Wins CRISPR Patent Battle." MIT Technology Review. February 15. https://www.technologyreview.com/s/603662/patent -office-hands-win-in-crispr-battle-to-broad-institute.

Regalado, Antonio. 2017b. "Semi-Synthetic Life Form Now Fully Armed and Operational." MIT Technology Review. November 29. https://www.technologyreview.com /s/609567/semi-synthetic-life-form-now-fully-armed-and-operational.

Resta, Robert. 2002. "Historical Aspects of Genetic Counseling: Why Was Maternal Age 35 Chosen as the Cut-off for Offering Amniocentesis?" *Medicina nei secoli* 14 (3): 793–811.

Resta, Robert. 2014. "NIPS SPIN." *The DNA Exchange* (blog), April 22. https://thedna exchange.com/2014/04/21/nips-spin.

Resta, Robert. 2018. "Work Shift: A (Wrong?) Prediction." *The DNA Exchange* (blog), June 4. https://thednaexchange.com/2018/06/04/work-shift-a-wrong-prediction.

Revive & Restore. 2018a. "How Genetic Rescue Works." http://reviverestore.org/what -we-do/extinction-continuum.

Revive & Restore. 2018b. "TedX De-extinction." http://reviverestore.org/events/ted xdeextinction/about.

Revive & Restore. 2018c. "What We Do." January 10. https://web.archive.org/web/20180314051621/http://reviverestore.org/what-we-do.

Rogers, Dale Evans. 1953. *Angel Unaware: A Touching Story of Love and Loss.* Westwood, NJ: F. H. Revell Co.

Romesberg, Floyd. 2016. "Expanding the Genetic Alphabet." TEDMED, September 27. https://www.tedmed.com/talks/show?id=529436.

Roosth, Sophia. 2017. *Synthetic: How Life Got Made.* 1st ed. Chicago, London: University of Chicago Press.

Rosenwald, Michael S., and Theresa Vargas. 2013. "Federal Probe Underway of Police-Custody Death of Man with Down Syndrome." *Washington Post,* sec. Local, July 19. https://www.washingtonpost.com/local/federal-probe-underway-of-police-custody-death-of-man-with-down-syndrome/2013/07/19/6480c064-f0b2-11e2-a1f9-ea873b7e0424_story.html.

Roslin Institute. 2018. "The Life of Dolly" (web page). Accessed April 19, 2018. http://dolly.roslin.ed.ac.uk/facts/the-life-of-dolly/index.html.

Rothman, Barbara Katz. 1993. *The Tentative Pregnancy: How Amniocentesis Changes the Experience of Motherhood.* 1993 edition. New York: Norton.

SafBaby. 2013. "Forest Kindergarten—A Better Way to Teach Our Young Children?" *SafBaby* (blog), August 6. https://safbaby.com/forest-kindergarten-a-better-way-to-teach-our-young-children.

Sample, Ian. 2010. "Craig Venter Creates Synthetic Life Form." *The Guardian,* May 20. http://www.theguardian.com/science/2010/may/20/craig-venter-synthetic-life-form.

Sandel, Michael J. 2007. *The Case against Perfection: Ethics in the Age of Genetic Engineering.* Cambridge, MA: Belknap Press of Harvard University Press.

Santoro, Stephanie L., Anna J. Esbensen, Robert J. Hopkin, Lesly Hendershot, Francis Hickey, and Bonnie Patterson. 2016. "Contributions to Racial Disparity in Mortality among Children with Down Syndrome." *The Journal of Pediatrics* 174. doi: 10.1016/j.jpeds.2016.03.023.

Saxton, Marsha. 2017. "Disability Rights Meets DNA Research." Filmed October 3 in St. Peter, MN. Nobel Conference 53: Reproductive Technology: How Far Do We Go? https://www.youtube.com/watch?v=6I3OR4GvJrU&list=PLHuAoPzfQhGExR6bKzAsdoyOWTjQ8Pxvj&index=6&t=0s.

Scheufele, Dietram. 2016. CRISPR: The Promise and the Peril. https://grow.cals.wisc.edu/health/crispr-the-promise-and-the-peril.

Securities and Exchange Commission. 2011. "SEC ADMINISTRATIVE PROCEEDING File No. 3-14524." September 1. https://www.sec.gov/litigation/admin/2011/34-65247.pdf.

Securities and Exchange Commission. 2010. "SEC Charges Former Biotech Company Executive for False Claims about Down Syndrome Test," 2010-94. Press release, June 2. https://www.sec.gov/news/press/2010/2010-94.htm.

Selden, Steven. 1999. *Inheriting Shame: The Story of Eugenics and Racism in America.* New York: Teachers College Press.

Sequenom. 2013. Home page. January 13. https://web.archive.org/web/20130113 044726/http://www.sequenom.com/

Sequenom. 2014a. Home page. June 25. https://web.archive.org/web/20140625 094014/http://www.sequenom.com.

Sequenom. 2014b. "Sequenom Laboratories—About Us." September 25. https://web .archive.org/web/20140925095516/http://laboratories.sequenom.com:80/about-us /sequenom-laboratories-only-results-matter-are-accurate-ones.

Sequenom. 2017. "MaterniT21® PLUS." September 5. https://web.archive.org/web /20170905153137/https://www.sequenom.com/tests/reproductive-health/mater nit21-plus#patient-overview.

Sequenom. 2018a. "Down Syndrome Testing—What You Need to Know." January 9. https://web.archive.org/web/20180109033326/https://www.sequenom.com/down -syndrome-testing.

Sequenom. 2018b. Home page. July 1. https://web.archive.org/web/20180701173758 /https://www.sequenom.com/#.

Sequenom. 2018c. "MaterniT21 Brochure." January 9. https://web.archive.org/web /20180109050320/https://www.sequenom.com/uploads/collateral/31-20578R1 .0_MaterniT_21_PLUS_ESS_patientbrochure.pdf.

Shapiro, Beth. 2016. *How to Clone a Mammoth: The Science of De-Extinction.* Reprint edition. Princeton, NJ: Princeton University Press.

Shipman, Seth L., Jeff Nivala, Jeffrey D. Macklis, and George M. Church. 2017. CRISPR–Cas Encoding of a Digital Movie into the Genomes of a Population of Living Bacteria. *Nature* 547 (7663): 345. doi:10.1038/nature23017.

Shriver, Eunice Kennedy. 1962. "Hope for Retarded Children." *The Saturday Evening Post,* September 22. http://www.eunicekennedyshriver.org/articles/article/148.

Silver, Lee M. 1997. *Remaking Eden: How Genetic Engineering and Cloning Will Transform the American Family.* 1st ed. New York: Harper Perennial.

Silver, Lee M. 2013. Method and system for generating a virtual progeny genome. US8620594 B2, filed August 22, 2012, and issued December 31, 2013. https://www .google.com/patents/US8620594.

Solnit, Rebecca. 2004. *River of Shadows: Eadweard Muybridge and the Technological Wild West*. Reprint edition. New York: Penguin.

Solnit, Rebecca. 2013. *The Faraway Nearby*. New York: Penguin.

Solomon, Andrew. 2012. *Far From the Tree: Parents, Children and the Search for Identity*. New York: Scribner.

Solomon, Andrew. 2016. "Literature about Medicine May Be All That Can Save Us." *The Guardian* (US edition), sec. Books, April 22. http://www.theguardian.com/books/2016/apr/22/literature-about-medicine-may-be-all-that-can-save-us.

Stephens, John Franklin. 2012. "An Open Letter to Ann Coulter." *Huffington Post* (blog), October 25. https://www.huffingtonpost.com/timothy-p-shriver/an-open-letter-to-ann-coulter_b_2012454.html.

Stephens, John Franklin. 2016. "How Bad Is Gary Owen's Comedy Routine on Showtime?" *Huffington Post* (blog), April 26. https://www.huffingtonpost.com/john-franklin-stephens/how-bad-is-gary-owens-com_b_9780334.html.

Stern, Alexandra Minna. 2002. "Making Better Babies: Public Health and Race Betterment in Indiana, 1920–1935." *American Journal of Public Health* 92 (5): 742–752.

Stern, Alexandra Minna. 2012. *Telling Genes: The Story of Genetic Counseling in America*. 1st ed. Baltimore: Johns Hopkins University Press.

Stern, Alexandra Minna. 2015. *Eugenic Nation: Faults and Frontiers of Better Breeding in Modern America*. 2nd ed. Oakland: University of California Press.

Stevens, Hallam. 2016. *Biotechnology and Society: An Introduction. Reprint edition*. Chicago: University of Chicago Press.

Stoll, Katie A. 2014. "NIPS: Microdeletions, Macro Questions." *The DNA Exchange* (blog), November 2. https://thednaexchange.com/2014/11/02/guest-post-nips-micro deletions-macro-questions.

Thompson, Charis. 2017. "The End of the World As We Know It? Human Technology Futures in a Time of Automation, Augmentation, and Deselection." Filmed October 4 in St. Peter, MN. Nobel Conference 53: Reproductive Technology: How Far Do We Go? https://www.youtube.com/watch?v=L3VHy_Mfk7Y&list=PLHuAo PzfQhGExR6bKzAsdoyOWTjQ8Pxvj&index=11.

Tingley, Kim. 2014. "The Brave New World of Three-Parent I.V.F." *New York Times*, June 27. https://www.nytimes.com/2014/06/29/magazine/the-brave-new-world-of-three-parent-ivf.html?_r=0.

Toynbee, Polly. 2014. "This Treatment Would Save Children's Lives—So Why Won't the Government Allow It?" *The Guardian*, sec. Opinion, February 11. https://

www.theguardian.com/commentisfree/2014/feb/11/treatment-save-children-lives
-mitochondrial-replacement.

Trent, James. 2016. *Inventing the Feeble Mind: A History of Intellectual Disability in the United States*. 1st ed. Oxford: Oxford University Press.

Vargas, Theresa. 2013. "Questions Haunt Family of Man with Down Syndrome Who Died in Police Custody." *Washington Post*, sec. Local, July 18. https://www
.washingtonpost.com/local/questions-haunt-family-of-man-with-down-syndrome
-who-died-in-police-custody/2013/07/18/693d4c2a-eef6-11e2-bed3-b9b6fe264871
_story.html.

Vargas, Theresa. 2018. "Settlement Reached in Police-Custody Death of Man with Down Syndrome." *Washington Post*, sec. Local, April 24. https://www.washingtonpost
.com/local/settlement-reached-in-police-custody-death-of-man-with-down-syn
drome/2018/04/24/7d53c0ca-47fe-11e8-827e-190efaf1f1ee_story.html.

Venter, J. Craig. 2014. *Life at the Speed of Light: From the Double Helix to the Dawn of Digital Life*. Reprint edition. New York: Penguin.

Verifi. 2014. Home page. August 27. https://web.archive.org/web/20140827110633
/http://www.verifitest.com.

Vogell, Heather. 2015. "Unrestrained." Text/html. ProPublica. December 10. https://
www.propublica.org/article/advoserv-profit-and-abuse-at-homes-for-the-profoundly
-disabled.

Wade, Nicholas. 2010. "Synthetic Bacterial Genome Takes Over Cell." *New York Times*, May 20, sec. Science. https://www.nytimes.com/2010/05/21/science/21cell.html.

Watson, James D., and Francis H. Crick. 1953. "Molecular Structure of Nucleic Acids: A Structure for Deoxyribose Nucleic Acid." *Nature* 171 (April): 737–738.

Weaver, Matthew. 2015. "Dyslexic Donors Turned Away from Largest UK Sperm Bank." *The Guardian*, sec. News, December 29. https://www.theguardian.com/soci
ety/2015/dec/29/largest-uk-sperm-bank-turns-away-dyslexic-donors.

Weintraub, Karen. 2016. "'3-Parent Baby' Procedure Faces New Hurdle." *Scientific American*, November 30. https://www.scientificamerican.com/article/ldquo-three-parent
-baby-rdquo-procedure-faces-new-hurdle.

Wexler, Alice. 1996. *Mapping Fate: A Memoir of Family, Risk, and Genetic Research*. Berkeley: University of California Press.

White, Raymond E. 2006. *King of the Cowboys, Queen of the West: Roy Rogers and Dale Evans*. Madison, WI: Popular Press.

Williams, Patricia. 2015. "Who's Getting Rich off Your Genes?" *The Nation*, April 3. https://www.thenation.com/article/whos-getting-rich-your-genes.

Williams, William Carlos. 1944. *The Collected Poems: Volume II, 1939–1962*. New York: New Directions.

Wissink, Inge B., Eveline van Vugt, Xavier Moonen, Geert-Jan J. M. Stams, and Jan Hendriks. 2015. "Sexual Abuse Involving Children with an Intellectual Disability: A Narrative Review." *Research in Developmental Disabilities* 36 (Jan.): 20–35.

Wolpe, Paul Root. 2010. *It's Time to Question Bio-Engineering*. TED. https://www.ted .com/talks/paul_root_wolpe_it_s_time_to_question_bio_engineering.

Wu, Tim. 2017. *The Attention Merchants: The Epic Scramble to Get Inside Our Heads*. Reprint edition. New York: Vintage.

Wynne, Brian. 2001. "Public Understanding of Science." In *Handbook of Science and Technology Studies*. Revised edition, ed. Sheila Jasanoff, Gerald E. Markle, James C. Peterson and Trevor J. Pinch, 361–388. Thousand Oaks, CA: Sage.

Yamada, Mitsutoshi, Valentina Emmanuele, Maria J. Sanchez-Quintero, Bruce Sun, Gregory Lallos, Daniel Paull, Matthew Zimmer, et al. 2016. "Genetic Drift Can Compromise Mitochondrial Replacement by Nuclear Transfer in Human Oocytes." *Cell Stem Cell* 18 (6): 749–754. doi:10.1016/j.stem.2016.04.001.

Yong, Ed. 2013. "Shutting Down the Extra Chromosome in Down's Syndrome Cells." *Not Exactly Rocket Science* (blog), July 17. http://phenomena.nationalgeographic.com /2013/07/17/how-to-shut-down-the-extra-chromosome-in-downs-syndrome.

Yong, Ed. 2017. "Scientists Can Use CRISPR to Store Images and Movies in Bacteria." *The Atlantic,* July 12. https://www.theatlantic.com/science/archive/2017/07/scien tists-can-use-crispr-to-store-images-and-movies-in-bacteria/533400.

Young, Stella. 2014. "I'm Not Your Inspiration, Thank You Very Much." Filmed in April in Sydney. TedX Sydney, 9:13, https://www.ted.com/talks/stella_young _i_m_not_your_inspiration_thank_you_very_much.

Zentner, Gabriel E., and Michael J. Wade. 2017. "The Promise and Peril of CRISPR Gene Drives." *BioEssays* 39 (10). https://doi.org/10.1002/bies.201700109.

Index